Prevalence, Fate and Effects of Plastic in Freshwater Environments

Prevalence, Fate and Effects of Plastic in Freshwater Environments

Editor

Farhan R. Khan

MDPI • Basel • Beijing • Wuhan • Barcelona • Belgrade • Manchester • Tokyo • Cluj • Tianjin

Editor
Farhan R. Khan
Roskilde University
Denmark
University of Oslo
Norway

Editorial Office
MDPI
St. Alban-Anlage 66
4052 Basel, Switzerland

This is a reprint of articles from the Special Issue published online in the open access journal *Toxics* (ISSN 2305-6304) (available at: https://www.mdpi.com/journal/toxics/special_issues/plastic_freshwater).

For citation purposes, cite each article independently as indicated on the article page online and as indicated below:

LastName, A.A.; LastName, B.B.; LastName, C.C. Article Title. *Journal Name* **Year**, *Volume Number*, Page Range.

ISBN 978-3-0365-1297-6 (Hbk)
ISBN 978-3-0365-1298-3 (PDF)

Contents

About the Editor

Farhan R. Khan (Dr) obtained his Ph.D.in Environmental Science from King's College London, studying the ecotoxicology and ecophysiology of trace metals in the aquatic environment. Since then, his research interests have extended to include anthropogenic particles, namely engineered nanoparticles, micro- and nano-plastics, and car-tire wear particles. The determination of the occurrence of plastics in less-researched locations, notably well-known freshwaters such as Lake Victoria and the River Nile, has been a particular focus, as well as furthering our understanding of pollutant bioavailability and the toxicological risk resulting from microplastic exposure.

Editorial

Prevalence, Fate and Effects of Plastic in Freshwater Environments: New Findings and Next Steps

Farhan R. Khan

Department of Science and Environment, Roskilde University, Universitetsvej 1, PO Box 260, DK-4000 Roskilde, Denmark; frkhan@ruc.dk or farhan.khan@gmx.com; Tel.: +46-5019-0429

Received: 1 September 2020; Accepted: 9 September 2020; Published: 17 September 2020

At a time when a global pandemic rightly holds our collective attention, environmental issues have taken a backseat to the ongoing battle against Covid-19. However, when normality resumes (or a new normal emerges), these environmental concerns will return to the fore and few issues are as tangible or emotive as plastic pollution. The presence of plastics in aquatic environments, particularly freshwater environments that are relied upon on a daily basis (e.g., for water and food supply, as well as recreation and tourism), is a constant reminder of the indelible mark that our species is leaving on the planet. Indeed, whilst plastic pollution has been described as a potential hallmark of the Anthropocene [1], our understanding of its presence and impacts on freshwater systems still lags far behind the corresponding knowledge that has been amassed from marine (coastal and oceanic) locales [2–4]. In this Special Issue on the "Prevalence, Fate and Effects of Plastic in Freshwater Environments", original research articles report the latest findings describing the occurrence of plastics and microplastics (MPs, <5 mm in size) within previously uninvestigated rivers and lakes. As important as mapping plastics pollution across global freshwaters is gaining an understanding into the effects of the interactions between MPs, formed from the degradation of larger debris (secondary MPs) or purposely manufactured microscopic particles (primary MPs), and biota. Laboratory-based mechanistic studies examined the effects of MP exposure to freshwater organisms, with endpoints ranging from cellular to behavioral. This Editorial aims to highlight the papers presented within the Special Issue and place them into the broader context of freshwater plastics research.

To date, few studies reporting plastic and MP prevalence in freshwaters have originated from the African continent [4], despite the presence of some of the world's most notable lakes and rivers. This knowledge gap needs to be urgently addressed [5] and thus it was especially pleasing that research from Lake Malawi and the River Nile contributed to this Special Issue. Focusing on macro-sized anthropogenic litter, the study by Mayoma et al. [6] resulted from an NGO-led citizen science project to clean up the shorelines of Lake Malawi. Over 2000 volunteers participated in four annual beach clean-ups from 2015 to 2018. Almost half a million individual items of beach debris were collected, of which approximately 80% was plastic. Glass, metal and paper were also found at each site, but plastic litter was by far the most abundant. The composition of plastic litter varied between sites, but carrier bags (22.5–49.4%), personal hygiene products (7.15–18.6%), beverage bottles (1.9–18.8%) and discarded fishing gear (0.21–20.25%) represented the major categories. The study highlighted the important role that volunteers could play in the remediation of the environment and the collection of scientific data, as has been discussed elsewhere for freshwater systems [7]. Moreover, by focusing on macroplastics in the African Great Lakes, the study provides a link between the use (misuse) of plastic products on the shoreline and the ingestion of MPs by resident lake species, as has been noted elsewhere in the region (Lake Victoria, [8]).

The research conducted in Lake Victoria was the first to describe the ingestion of MPs by African freshwater species (Nile perch and Nile tilipia) [8] and, although subsequent similar research from different parts of the continent has emerged, documenting the occurrence of MPs in the Nile River, the world's longest river, and its biota has been a surprising knowledge gap. The Nile River flows from the heart of Africa to the Mediterranean Sea, but the presence of MPs was determined in Cairo, Egypt's capital, with an estimated population of 20 million people. Analyzing the digestive tracts of Nile tilapia (*Oreochromis niloticus*) and catfish (*Bagrus bayad*) revealed the presence of MPs in over 75% of the sampled fish [9]. Similar studies from both marine and freshwater environments showed that this high level of MP ingestion is rarely found and that fish sampled from the Nile River in Cairo are potentially among the most in danger of consuming MPs worldwide. The gastrointestinal tracts of omnivorous tilipia contained significantly more MPs than those of the piscivorous catfish, which is likely due to the former's more varied diet and increased potential to mistake plastic items for food. In keeping with similar studies, fibers were the most abundant MP type, followed by films and fragments. The diverse origins of fibers that may result from the degradation of clothing, furniture or fishing gear means that it is difficult to pinpoint any one source of MPs beyond general urbanization. Polyethylene (PE), polyethylene terephthalate (PET) and polypropylene (PP) were all identified, and these polymers are widely used in a plethora of plastic products. The first detailing of MPs in fish from the Nile River revealed the scale of the problem, but further research is needed understand and mitigate its effects.

As famous as the Nile River is the Mississippi River (USA). During periods of flooding along the lower Mississippi River, its waters are diverted through the Bonnet Carré Spillway into the coastal area of the Mississippi Sound, which prevents flooding downstream in the city of New Orleans. Such a major flooding event occurred during 2019. The study by Scircle et al. (2020) [10] captured the effect of this on the presence of MPs in the waters housing oyster (*Crassostrea virginica*) reefs in the Mississippi Sound. Although coastal, during the spring and summer of the 12 month sampling campaign, historic flooding of the Mississippi River caused major freshwater intrusions into the sampled locations. Across the ten sample sites and seasonal sampling periods, there was considerable spatial and temporal variation, as might be expected, with the average MP concentration ranging from 30 to 192 MPs/L across sites. During the period of freshwater intrusions, MP abundances tended to correlate with salinity. In other words, sites most impacted by freshwater intrusions had the least MPs owing to a dilution effect, but this effect was temporary. There were no significant changes in the relative distribution of MPs during freshwater intrusions, with most of the MPs (>50%) in the lower size fraction (~25–90 μm). Fragments (~84%), fibers (~11%) and beads (~5%) constituted the MP types. Polyester, acrylates/polyurethanes, polyamide, polypropylene, polyethylene and polyacetal were all identified polymers. This work provides empirical data on the waters of the Mississippi Sound, in which oyster reefs are located. More research is needed to determine the impacts on the organisms. However, this investigation importantly demonstrates how MP pollution is impacted by changing conditions and thus represents a counterpoint to the numerous studies that only capture MP pollution during a 'single snapshot in time'. Temporal and seasonal events need greater recognition.

The study from the Nile River [9] illustrates, again, if further proof was necessary, the already established phenomena of fish indiscriminately ingesting MPs, but the impacts on the fish are not fully understood. Inflammatory responses and intestinal health by way of alterations of the microbiome resulting from MP exposure constitute one such area that requires further elucidation. In the study by Kurchaba et al. (2020) [11], larval zebrafish (5 days post fertilization) were exposed to PE MPS for four or ten days. Markers of metabolic disturbance and inflammation were assessed through physiological and genomic analysis. The ten-day exposure to 20 mg MPs/L did not seem to affect the overall metabolism of the larval zebrafish and had limited impact on inflammatory molecular responses. However, it appeared to trigger an elevated reactive oxygen species (ROS) response, with a significant increase in the oxidative stress mediator L-FABP (liver fatty-acid binding protein). This increase in ROS was accompanied by a higher abundance of *Bacteriodetes*, and although further research is needed to verify the link between these two outcomes, the authors speculated that increased oxidative stress in MP fish promotes the growth of these bacteria. Localization of ROS and concurrent dysbiosis of the larval microbiome indicate that aquatic organisms are negatively affected by MP exposure, which may render the animal more susceptible to diseases. The gut microbiome has also emerged as a new growth area in explaining adverse human health and disease. The finding that MP exposure disrupts the microbiome of zebrafish may have important ramifications for humans, especially since it is known that MPs have been found in a variety of foodstuffs.

The potential for oxidative damage following PE MP exposure was also assessed by Scopetani et al. (2020) [12] using the freshwater oligochaete *Tubifex tubifex*. Worms were exposed via water or sediment separately or simultaneously spiked with MPs. Mortality was assessed over 120 h and oxidative stress status was assessed via enzymatic assays for the biomarkers glutathione reductase and peroxidase. At concentrations used (2 g/L (*w/v*) in water and 2 mg/g (*w/w*) in sediment), *Tubifex* survival was not significantly affected. Similarly, there was no effect on the activities of either biomarker. Other species have been affected by MPs at similar concentrations and thus the results described in this study may be due to tolerance of the chosen test species. Importantly, though, the lack of effects should not diminish the findings and, as in the case of all new pollutants (including MPs), it is necessary that negative results are also made available so that associated risks may be discussed with nuance.

The use of biomarkers has been widely debated in ecotoxicology, with one area of concern being the link between intracellular responses and physiological or behavioral response. In the study by Pflugmacher et al. (2020) [13], the annelid *Enchytraeus crypticus* was subject to a choice experiment in which different amounts of high-density PE MPs taken from green bottle caps were mixed into soil. In each test, worms chose soil without MPs or the area with a lower MP concentration. Contact with MPs resulted in enhanced oxidative stress measured through the antioxidative enzymes catalase and glutathione S-transferase. Explaining the results, the authors noted that the exposure time was too short for chemicals to leach from the plastics and that the MPs were too large to be ingested. Thus, the likely cause of avoidance behavior was that the addition of MPs adversely changed the properties of the soil.

In summary, this collection of original research articles provides a valuable update of recent investigations into the global reach of plastic pollution within freshwaters and the biological effects induced by its presence. The question of how freshwater plastics research should proceed is of paramount importance, but there are no easy 'next steps'. Clearly, more information is needed on the pervasive presence and effects of plastics in freshwaters, pursuing monitoring studies that include temporal and spatial factors that affect, for instance, hydrology and water flow and also accounting for a wider range of localized and systemic endpoints. As the studies within this Special Issues illustrate, there is broad range of topics to cover, and whilst calls have been made to harmonize and standardize plastics research, this belies the realities of working in such a diverse field and particularly the challenges in conducting scientific inquiry in different parts of the world. Approaches taken within each study must therefore endeavor to match the research aims of that singular investigation whilst simultaneously moving the research area towards a more robust and dynamic framework that can meet the future challenges of an ever-expanding group of anthropogenic particulates, including, but not limited to, the emergent concerns surrounding nanoplastics [14] and tire wear particles [15]. Moreover, returning to the opening remarks of this Editorial on the current pandemic, the outbreak of Covid-19 has increased the use and discard of personal protective equipment, which heightens the need to focus on plastic waste generation, its impacts and mitigation [16,17]; from its ecological and ecotoxicological impacts through to defining improved strategies for waste management. The studies presented in this Special Issue add to the weight of evidence that recognizes plastic pollution as a significant freshwater problem that requires immediate attention. The consequences of inaction are severe.

Funding: This research received no external funding.

Conflicts of Interest: The author declares no conflict of interest.

References

1. Zalasiewicz, J.; Waters, C.N.; Do Sul, J.A.I.; Corcoran, P.L.; Barnosky, A.D.; Cearreta, A.; Edgeworth, M.; Gałuszka, A.; Jeandel, C.; Leinfelder, R.; et al. The geological cycle of plastics and their use as a stratigraphic indicator of the Anthropocene. *Anthropocene* **2016**, *13*, 4–17. [CrossRef]
2. Wagner, M.; Scherer, C.; Alvarez-Munoz, D.; Brennholt, N.; Bourrain, X.; Buchinger, S.; Fries, E.; Grosbois, C.; Klasmeier, J.; Marti, T.; et al. Microplastics in freshwater ecosystems: What we know and what we need to know. *Environ. Sci. Eur.* **2014**, *26*, 12. [CrossRef]
3. Eerkes-Medrano, D.; Thompson, R.; Aldridge, D. Microplastics in freshwater systems: A review of the emerging threats, identification of knowledge gaps and prioritisation of research needs. *Water Res.* **2015**, *75*, 63–82. [CrossRef] [PubMed]
4. Blettler, M.C.M.; Abrial, E.; Khan, F.R.; Sivri, N.; Espinola, L.A. Freshwater plastic pollution: Recognizing research biases and identifying knowledge gaps. *Water Res.* **2018**, *143*, 416–424. [CrossRef] [PubMed]
5. Khan, F.R.; Mayoma, B.S.; Biginagwa, F.J.; Syberg, K. Microplastics in Inland African Waters: Presence, Sources, and Fate. In *Freshwater Microplastics the Handbook of Environmental Chemistry*; Wagner, M., Lambert, S., Eds.; Springer: Berlin/Heidelberg, Germany, 2018; Volume 58, pp. 101–124.
6. Mayoma, B.S.; Mjumira, I.S.; Efudala, A.; Syberg, K.; Khan, F.R. Collection of Anthropogenic Litter from the Shores of Lake Malawi: Characterization of Plastic Debris and the Implications of Public Involvement in the African Great Lakes. *Toxics* **2019**, *7*, 64. [CrossRef] [PubMed]
7. Syberg, K.; Hansen, S.F.; Christensen, T.B.; Khan, F.R. Risk perception of plastic pollution: Importance of stakeholder involvement and citizen science. In *Freshwater Microplastics the Handbook of Environmental Chemistry*; Wagner, M., Lambert, S., Eds.; Springer: Berlin/Heidelberg, Germany, 2018; Volume 58, pp. 203–221.
8. Biginagwa, F.; Mayoma, B.; Shashoua, Y.; Syberg, K.; Khan, F.R. First evidence of microplastics in the African Great Lakes: Recovery from Lake Victoria Nile perch and Nile tilapia. *J. Great Lakes Res.* **2016**, *42*, 1146–1149. [CrossRef]
9. Khan, F.R.; Shashoua, Y.; Crawford, A.; Drury, A.; Sheppard, K.; Stewart, K.; Sculthorp, T. 'The Plastic Nile': First Evidence of Microplastic Contamination in Fish from the Nile River (Cairo, Egypt). *Toxics* **2020**, *8*, 22. [CrossRef] [PubMed]
10. Kurchaba, N.; Cassone, B.J.; Northam, C.; Ardelli, B.F.; LeMoine, C.M.R. Effects of MP Polyethylene Microparticles on Microbiome and Inflammatory Response of Larval Zebrafish. *Toxics* **2020**, *8*, 55. [CrossRef] [PubMed]
11. Scircle, A.; Cizdziel, J.V.; Tisinger, L.; Anumol, T.; Robey, D. Occurrence of Microplastic Pollution at Oyster Reefs and Other Coastal Sites in the Mississippi Sound, USA: Impacts of Freshwater Inflows from Flooding. *Toxics* **2020**, *8*, 35. [CrossRef] [PubMed]
12. Scopetani, C.; Esterhuizen, M.; Cincinelli, A.; Pflugmacher, S. Microplastics Exposure Causes Negligible Effects on the Oxidative Response Enzymes Glutathione Reductase and Peroxidase in the Oligochaete *Tubifex tubifex*. *Toxics* **2020**, *8*, 14. [CrossRef] [PubMed]
13. Pflugmacher, S.; Huttunen, J.H.; Von Wolff, M.A.; Penttinen, O.P.; Kim, Y.J.; Kim, S.; Mitrovic, S.M.; Esterhuizen-Londt, M. *Enchytraeus crypticus* Avoid Soil Spiked with Microplastic. *Toxics* **2020**, *8*, 10. [CrossRef] [PubMed]
14. Da Costa, J.P.; Santos, P.S.; Duarte, A.C.; Rocha-Santos, T. (Nano) plastics in the environment–sources, fates and effects. *Sci. Total Environ.* **2016**, *566*, 15–26. [CrossRef] [PubMed]
15. Halle, L.L.; Palmqvist, A.; Kampmen, K.; Khan, F.R. Ecotoxicology of micronized tire rubber: Past, present and future considerations. *Sci. Total Environ.* **2020**, *706*, 135694. [CrossRef] [PubMed]
16. Prata, J.C.; Silva, A.L.; Walker, T.R.; Duarte, A.C.; Rocha-Santos, T. COVID-19 pandemic repercussions on the use and management of plastics. *Environ. Sci. Technol.* **2020**, *54*, 7760–7765. [CrossRef] [PubMed]
17. Singh, N.; Tang, Y.; Ogunseitan, O.A. Environmentally sustainable management of used personal protective equipment. *Environ. Sci. Technol.* **2020**, *54*, 8500–8502. [CrossRef] [PubMed]

Article

Collection of Anthropogenic Litter from the Shores of Lake Malawi: Characterization of Plastic Debris and the Implications of Public Involvement in the African Great Lakes

Bahati S. Mayoma [1,*], Innocent S. Mjumira [2], Aubrery Efudala [2], Kristian Syberg [3] and Farhan R. Khan [3]

1 Department of Biology, University of Dodoma, Dodoma PO Box 338, 255, Tanzania
2 Beach and Underwater Cleanup Malawi, Lumbadzi PO Box 174, 265, Malawi; beach.underwatercleanupmalawi@gmail.com (I.S.M.); efudalaaubrey@gmail.com (A.E.)
3 Department of Science and Environment, Roskilde University, Universitetsvej 1, PO Box 260, DK-4000 Roskilde, Denmark; ksyberg@ruc.dk (K.S.); frkhan@ruc.dk (F.R.K.)
* Correspondence: bsosthenes@yahoo.com

Received: 26 November 2019; Accepted: 10 December 2019; Published: 13 December 2019

Abstract: Anthropogenic debris is an environmental problem that affects beaches and coastlines worldwide. The abundance of beach debris is often documented with the use of public volunteers. To date, such community participations have been largely confined to the marine environment, but the presence and impact of anthropogenic debris on freshwater shorelines has been increasingly recognized. Our study presents the first such information from the African Great Lakes, specifically Lake Malawi. A total of 490,064 items of anthropogenic litter were collected by over 2000 volunteers in a clean-up campaign that took place annually between 2015 and 2018. Approximately 80% of the anthropogenic debris was comprised of plastic litter, with plastic carrier bags being the most common item. The dominance of plastic litter, and in particular the presence of plastic bags, which have subjected to bans in some African countries, is discussed. The broader implications of citizen science in the African Great Lakes area is also discussed.

Keywords: plastics; plastic debris; African great lakes; freshwater; beach clean-up; citizen science

---

1. Introduction

Anthropogenic debris is an environmental problem that affects beaches and coastlines worldwide [1,2]. Typically consisting of larger items (> 2 cm) of plastics, glass, metal, paper, and wood, plastic litter is often recorded as the most abundant type [3,4]. Although the presence of litter may decrease the tourist potential of affected beaches [5], anthropogenic debris is not a purely aesthetic issue, as it also has ecological consequences and is detrimental to wildlife. Debris, particularly plastics, pose a risk of entanglement to aquatic animals, including fish and birds, and are also likely to be mistakenly ingested owing to their similarity to food items [6,7]. Plastics can also act as vectors transporting adsorbed chemical (e.g., hydrophobic persistent organic pollutants (POPs) or trace metals) or biological (e.g., microorganisms) loads into new geographical regions [8].

The documenting of beach debris has taken place globally on marine shorelines. Examples include clean-ups and surveys on beaches along the Pacific coast of Chile [3], in Greece on the Mediterranean Sea [9], in Orange County in California (USA, [10]), and a ten year nationwide assessment of anthropogenic litter on beaches of the British Isles [11], to name a few. These studies all have in common the utilization of volunteers in their collection of data during the beach survey and clean-up activities. In fact, over the last 10 years, the number of volunteers taking part in clean- ups has doubled from about half a million people in 2009 to just over a million in 2019 [12,13]. The use of volunteers allows for extensive sampling at a wider range of sites [14] and although data validity may be open to question, there is no statistical difference between the data obtained from volunteer clean-ups and those performed by experienced surveyors, provided a clear protocol is followed [15]. Furthermore, community participation and citizen science are considered to be effective methods of increasing public awareness of environmental challenges and has been particularly well used with plastics [16,17].

To date, such community participations have been largely confined to the marine environment, but the presence and impact of anthropogenic debris on freshwater shorelines has been increasingly recognized, with a handful of surveys and clean-ups conducted along the shoreline of the North American Great Lakes [18]. However, such information remains scarce for freshwater environments and has not yet been available for the African Great Lakes. This study presents, for the first time, data on the anthropogenic debris collected by volunteers on the Malawian coast of Lake Malawi. Lake Malawi is bordered by Malawi, Tanzania, and Mozambique in the East African rift valley region (Figure 1). Referred to as Lake Nyasa in Tanzania and Lake Niassa in Mozambique, Lake Malawi is the third largest lake on the African continent and the world's ninth largest lake overall (surface area: 29,500 km^2, mean depth: 264 m, maximum depth: 700 m, 19). Together with Lakes Victoria and Tanganyika, Lake Malawi holds approximately 25% of the world's total surface freshwater supply [19]. Lake Malawi is said to be the most species-rich lake in the world, with an estimated 500–1000 species of fish present in the lake [19]. Perhaps unsurprisingly, approximately 70% of the animal protein in the Malawian diet is in the form of fish [19]. The presence of debris, particularly plastics, on the shoreline of Lake Malawi could result in amounts of microplastics (MPs, < 5 mm) entering the water via degradation and being ingested by the fish. The ingestion of plastic particles by fish has been well documented globally from numerous aquatic habitats (e.g., [20–23]), including the African Great Lakes [24]. In this regard, understanding the land-based debris, as is the case with beach clean ups, may be an important step in understanding what enters the water.

In general, there is a scarcity of information regarding the presence and abundance of plastics (and microplastics) in Africa and specifically within the African Great Lakes [25], but two studies from Lake Victoria document the scope of the potential problem. Solid waste anthropogenic inputs were investigated in the Tanzanian waters of Lake Victoria by sampling across three main ecological zones; the nearshore (0–20 m depth), intermediate zone (20–40 m depth), and deep offshore waters (> 40 m) [26]. Plastic debris was found at all depths, with the dominant waste types originating from fishing activities; multifilament gillnets (44% of all debris), monofilament gillnets (42%), longlines and hooks (7%), and floats (1%). Plastic bags (4%) and clothing (2%) accounted for the remaining solid waste [26]. The second study was also conducted in the Tanzanian waters of Lake Victoria around the urban center of Mwanza and showed that two fish species, Nile perch (*Lates niloticus*) and Nile tilapia (*Oreochromis niloticus*), contained MPs within their digestive tracts [24]. In total, suspected plastic particles were recovered from the gastrointestinal tracts of 11 perch (55%) and 7 tilapia (35%), with spectral analysis confirming the presence of MPs in 20% of each species [24]. Together, these two studies provide evidence that plastic debris in Lake Victoria is subject to degradation and the products of that breakdown are available for ingestion by resident fish populations [25].

In the present, relatively small-scale, study, anthropogenic debris was surveyed and collected from three sites over a four-year period (2015–2018 inclusive) located in Central and Northern regions of Malawi. The clean-up sites represent the busiest beaches along Lake Malawi. In total, over 2000 volunteers took part in the clean-up activities, which were directed by Malawi beach and underwater clean-up, an environmental NGO. The protocol to instruct the volunteers was that of the ocean conservancy clean-up, which has been widely used in coastal locations. The data presented here adds to the little information on anthropogenic and plastic debris collected from freshwater shorelines and specifically is the first of its kind from the African Great Lakes.

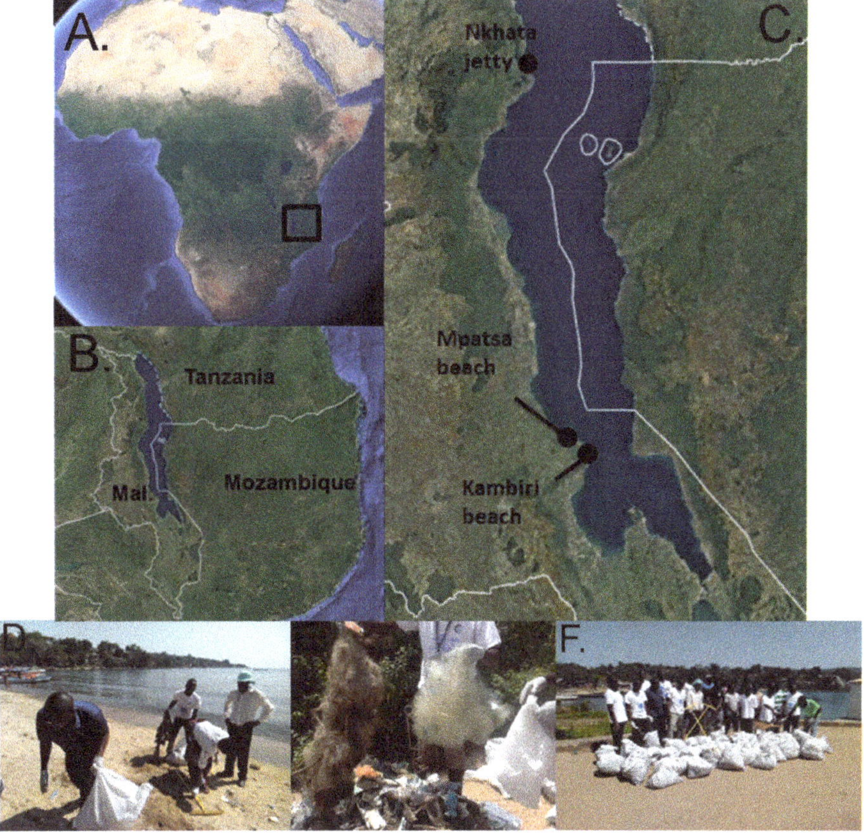

Figure 1. Location of the three beach clean-up sites on the Malawian coast of Lake Malawi (**A–C**, Mal. = Malawi). Images **D–F** show the volunteers during the clean-up and typical examples of the debris collected during the clean-up at Nkhata jetty in 2018.

2. Materials and Methods

2.1. Study Sites

The present study was conducted along the Malawian shorelines of Lake Malawi. Tourism, transportation, trade, agriculture, and fisheries related activities are found within a growing number of towns along the shoreline. Several district townships have hydrological linkage with the lake shoreline through rivers, stream, and storm waters. The most notable townships in Malawi within proximity of Lake Malawi (estimated population from 2018 census) are: Lilongwe (> 1,630,000), Mangochi (> 600,000), Salima (> 470,000), Karonga (> 365,000), and Nkhata bay (> 285,000). Three beaches were selected for clean-ups based on their proximity to urbanization, tourism, fishing, commerce, and transportation. Kambiri beach (34.617952°, –13.781318°) and Mpatsa beach (34.602600°, –13.762000°) are located towards the southern end of Lake Malawi, whereas Nkhata jetty (34.290731°, –11.606589°) is more northernly (Figure 1). Whilst the two beaches are favorite tourist destinations and also home to fishing and farming activities, Nkhata jetty forms a strategic port hub linking Malawi with the other countries that border Lake Malawi (Table 1). The port has various socio-economic activities, including construction, transport, and trade.

Table 1. Site descriptions.

Site	Location (Longitude, Latitude)	Description and Main Anthropogenic Activities
Kambiri beach	34.617952°, −13.781318°	Sandy beach with relatively gentle slope and scattered human settlements Tourism, maize farming, fishing, open market, livestock present, transportation mostly by wooden boats
Mpatsa beach	34.602600°, −13.762000°	Sandy beach with relatively gentle slope and scattered human settlements Tourism, fishing, maize farming, transportation mostly by wooden boats
Nkhata jetty	34.290731°, −11.606589°	Port area which forms a transportation hub linking Malawi, Tanzania, Mozambique and inland islands; historical memorial site which attracts tourists, recreation, cassava farming, fishing and market. Beach with minimal sand accumulation. Has various infrastructure developments including warehouse to cater for imports and exports services.

2.2. Beach Clean-Up Methodology

Beach clean-ups were conducted each September from 2015 to 2018 during the dry season. Clean-ups were organized by the 'beach and underwater clean-up', an NGO based in Lumbadzi, Malawi. The timing coincided with that of the ocean conservancy's international coastal clean-up day, an annual global event which takes place in September and October. Each year only one beach was selected, as follows: Kambiri beach in 2015 and 2016, Mpatsa beach in 2017, and Nkhata jetty in 2018 (Table 2).

Table 2. Details of beach clean-up campaigns at each site and year.

Site	Years	Approximated Area Covered (m^2)	No. of Volunteers	Total No. of Items Collected	Total No. of Plastic Items Collected
Kambiri beach	2015	100,000	350	89,442	68,836
	2016	160,000	821	177,087	132,666
Mpatsa beach	2017	20,000	64	19,238	16,361
Nkhata jetty	2018	40,000	904	204,297	171,092

Upon arrival, each volunteer was registered and provided with clean-up equipment such as gloves, collection bags, and gumboots. At each site, the name, location, time, distance, and number of volunteers were recorded in a pre-designed data collection form according to Lewis (2002) [27]. During clean-up, a stretch of beach measuring 100 m in length and 20 m in width was identified and cleaned before moving to the next stretch. Litter was collected from the shallow water to the highest water strandline. This was done for a maximum of four hours depending on weather conditions on the day and the number of volunteers. A total of 2139 volunteers took part in cleanup as follows: 350 and 821 people in Kambiri beach in 2015 and 2106, respectively, 64 people in Mpatsa beach in 2017, and 904 people in Nkhata jetty in 2018 (Table 2).

The methodology for clean-up events was adopted from the international coastal cleanup protocol [12,13]. The protocol constitutes 19 individual steps which were followed in three phases of the clean-up: (1) before clean-up, (2) during clean-up, and (3) immediately after clean-up. The steps included the identification of safe collection sites and coordinators, and training of volunteers before the clean-up; dealing with entangled or injured animals and filling in data cards during the clean-up; and recycling and disposal after clean-up. Additionally, the compilation of clean-up data was performed.

Litter larger than 2 cm was collected by hand, enumerated, and then classified into categories: glass, metal, paper, construction material, and plastic [28]. For easy sorting of recyclable trash, volunteers were encouraged to work in teams of five people. Each volunteer was given one trash bag designated for the major groups of recyclable waste (e.g., plastic bottles, glass, aluminum) which were sorted as they went and recorded on the data card by the group leader. Other litter categories not suitable for recycling, such as plastic carrier bags, cigarette butts, fishing gears, and personal hygiene products, were identified, counted, and collected for disposal. After the clean-up, all recyclable items were transported to the recycling centers while the rest of the anthropogenic litter was transported to designated waste sites.

2.3. Data Handling

Litter items in each major category (glass, metal, paper, construction material, and plastic) were compiled and totaled, as described in the ocean conservancy protocols. Plastic litter was further divided as carrier bags, personal hygiene products, beverage bottles, fishing gear, cups, packaging materials, bottle caps, lids, food wrappers, cutlery, cigarette butts, tires, and straws/stirrers.

3. Results and Discussion

3.1. Collection of Anthropogenic Litter

A total of 490,064 items of anthropogenic litter were collected by over 2000 volunteers at the three locations over the four-year period. With reference to the number of items collected, Nkhata jetty (>200,000 items) was the most impacted site, followed by Kambiri beach (2016 (> 175,000 items) and 2015 (approximately 90,000 items) and the Mpatsa beach (approximately 20,000 items) (Table 2). Typically, the numbers of items collected at different locations are normalized to the area of the site in order to make more valid comparisons of litter densities. The order of sites with regard to litter density is: Nkhata jetty (5.11 items/m^2) > Kambiri beach (2016, 1.11 items/m^2) > Mpatsa beach (0.96 items/m^2) > Kambiri beach (2015, 0.89 items/m^2). These densities are marginally lower than those reported elsewhere, 2 to 10 items/m^2 along the coast of Chile [3] and 2.3 to 6.3 items/m^2 reported along Brazilian coast [4], but typical densities reported globally range from 0–2–5 items/m^2 [3]. An average density of 45,000 items/m^2 was recorded from Japanese beaches [29], though it should be noted that the authors counted fragmented Styrofoam as individual items which are not usually counted individually in other studies.

The apparent greater levels of litter at Nkhata jetty beach may be attributed to its function as a busy port which links Malawi to Tanzania. Ports and associated activities have been reported to cause high level of debris in their surrounding environment [30]. Conversely, Kambiri beach and Mpatsa beach have relatively little industrial usage, instead being utilized for tourism, fishing, and farming. However, such results need to be interpreted with caution, as the number of items found at each site was significantly correlated to the number of volunteers conducting the clean-up. Correlating the number of volunteers to total litter found (values presented in Table 2) resulted in an R^2 value of 0.995 (p = 0.00235, R program). Thus, it is difficult to make conclusive comments regarding the relative abundance and density of anthropogenic debris at the different sites on the Lake Malawi shoreline when citizen participation appears to be such an influential factor, but certainly the greater the number of volunteers, the more litter is likely to be collected. Furthermore, in this case where volunteer numbers greatly influenced the number of items collected, we refrain from further discussing densities, instead using the percentages of items in each category for comparative purposes.

The composition of debris across all beaches was plastic 80.2 ± 5.0%, metal 9.5 ± 2.5%, glass 4.9 ± 3.8%, paper 4.6 ± 4.0%, and construction material 0.9 ± 1.5%. Moreover, the percentages of each category remained relatively consistent across all sites, with plastic litter being by far the most abundant type of anthropogenic debris (Figure 2). Again, these results are in keeping with studies from other locations. European, North, and South American clean-ups have also reported plastics as the main constituent of anthropogenic litter on coastal beaches [3,9–11]. Whilst little data is available for freshwater shorelines, volunteer beach clean-ups along the North American Great Lakes revealed that typically more than 80% of anthropogenic litter is comprised of plastics [18]. Our study adds data from the African Great Lakes to the global pattern of plastics as being the dominant litter type along freshwater and marine shorelines.

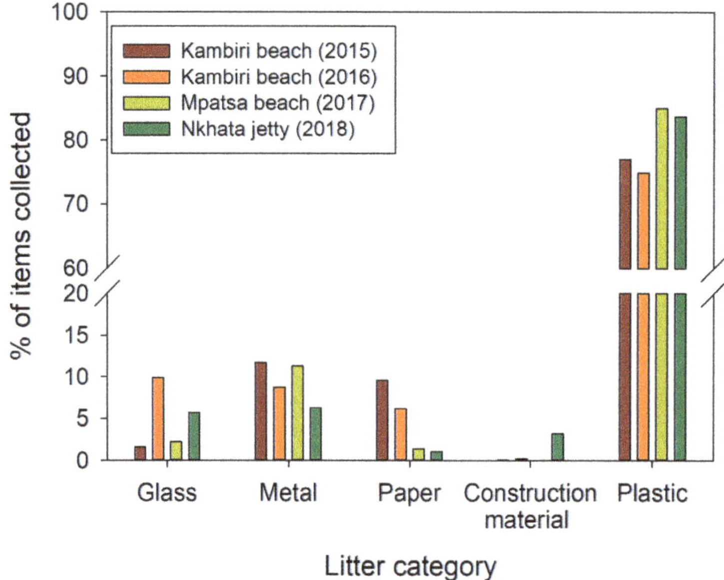

Figure 2. Percentage of items found of different categories of anthropogenic material (glass, metal, paper, construction material, and plastic) across sites and clean-up years.

3.2. Plastic Litter

As with anthropogenic debris, in general the composition of plastic debris was also relatively consistent across sites: plastic carrier bags (22.5–49.4%), personal hygiene products (7.15–18.6%), beverage bottles (1.9–18.8%), fishing gears (0.21–20.25%), cups (1.9–13.3%), packaging materials (1.3–13.8%), bottle caps (4.4–11.6%), lids (1.0–8.8%), and food wrappers (4.4–11.5%) (Figure 3). Remaining categories constituted less than 2% of plastic collected. The trend observed at Lake Malawi differs from that reported by the 2019 ocean conservancy annual report (13), which found that plastic bags are the sixth most common item found, whilst personal hygiene products (e.g., condoms, diapers, syringes), which rank second at Lake Malawi, are not within the top 10 items found on coastal beaches. In its annual clean-up report of 2010, the ocean conservancy [12] found that personal hygiene products contributed to only 1% of plastic items collected in African clean-ups.

Similar findings of plastic litter have also been reported elsewhere. The shoreline of the remote freshwater lake, Lake Hovsgol in northwest Mongolia, was dominated by plastics bags, beverage bottles, and discarded fishing gear [31]. Moreover, plastic bags constituted a small percentage of debris trawled from Lake Victoria [26] and the microplastic polymers found in the gastrointestinal tracts of fish sampled from Lake Victoria may have originated from this source [24], suggesting that shoreline waste may end up as the microplastics in the adjacent lake [25]. Lessons from Lake Victoria could be transferrable to Lake Malawi given the abundance of discarded plastic bags.

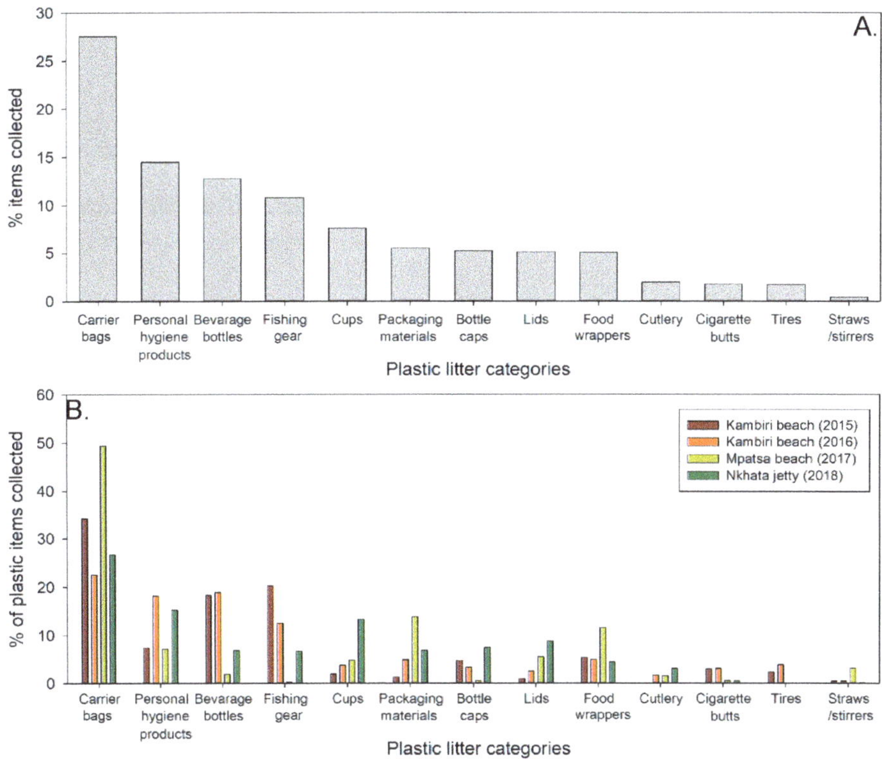

Figure 3. Breakdown of different categories of plastic debris as percentage of plastic items found at all sites and years combined (**A**) and by each site and year (**B**).

The issue of plastic bags is particularly prescient for Africa, as several countries have proceeded to reduce or ban plastic bag use. Such measures have been made on the grounds of environmental and public health, as discarded plastic bags have been shown to block gutters and drains which create storm water problems and collect water, which provides a breeding ground for mosquitos that spread malaria, and the use of bags as toilets has been linked to the spread of disease [32,33]. The government of South Africa introduced levies on plastic bag use in 2003 [34], and in Rwanda in 2005, a ban on the use and importation of plastic bags of < 100 microns thick was imposed [33]. Tanzania having made a similar ban based on thickness in 2006 has now, in June 2019, implemented a complete ban on manufacturing and importing plastic bags into country [35]. Developments in Tanzania may impact the amount of plastic waste entering Lake Malawi and may then also influence the policies of Malawi.

3.3. Implications for Public Involvement and Plastic Pollution in the African Great Lakes

The present study was the first to document the results of beach clean-up activities in the freshwater African Great Lakes. Though limited in size and scope, a pattern emerged that confirms previous coastal findings that plastic litter is the dominant type of anthropogenic debris. This study highlights the role that volunteers can play in both the remediation of the environment and the collection of scientific data. Citizen science has become a widely used initiative in combating plastic pollution. Organizations such as the national oceanic and atmospheric administration (NOAA) in the United States have developed a mobile phone application called "Marine Debris Tracker app" together with Southeast Atlantic marine debris initiative (SEA-MDI), which allows the public to report findings of litter from beaches and waterways [36]. Similarly, in the EU, the European environmental agency (EEA) has developed the "Marine LitterWatch" (MLW) program which collects beach litter data from both the monitoring efforts of the authorities and citizen science projects [37]. The MLW program provides scientific input to the EU policy process and thus illustrates the regulatory potential of these programs. The adoption of similar technologies may also aid the mapping of litter in other areas, and particularly in the African Great Lakes region, where information is scarce.

An interesting finding of this study is the strong positive correlation between volunteer numbers and total debris collected. This may suggest that not all litter was collected at each site, which may be due to the fixed time restrictions placed on the clean-up (4 h). Under these conditions, greater public involvement would lead to better results, but it should be noted that the number of sites is too low to make any conclusive statements on this relationship. One grouping that has been shown to be particularly valuable are students. Students can play an active role in collecting and monitoring data using mobile applications, such as the one made by NOAA. In just one example from the Roskilde Fjord region in Denmark, students collaborated with scientists to produce data on the occurrence of marine litter at 12 beaches around the fjord [17]. The students analyzed the data using a protocol inspired by the marine litter watch protocol and were shown to be able to follow instructions and generate reliable data [17]. The involvement of students in collecting data serves as an example of transformative learning [38] and helps to raise public awareness in general, and particularly in the next generation. Another advantage of involving students and schools is that by educating teachers on how to sample properly, they can ensure that the scientific protocol is properly followed.

Even though the described cases are from Europe and the United States, there is equal potential in other places, such as within Africa. It is not correct to say that African nations and people are unaware of the issues surrounding plastic pollution. Currently, Africa has the highest percentage of countries (~46%) with plastic bans [39] and numerous stakeholders are involving themselves in the fight against plastic pollution. Policy makers have enacted banning and reduction legislation and new technologies are being adopted across the plastics supply and recycling chain to enhance the possibilities of effective closed-loop waste management [40]. Clean-up activities, such as the one reported here from the shorelines of Lake Malawi, can be used to remove anthropogenic debris and raise public awareness.

Author Contributions: Conceptualization, B.S.M., I.S.M., A.U.,F.R.K.; methodology, I.S.M., A.U.; investigation, I.S.M., A.E.; data curation, B.S.M., I.S.M., A.E., F.R.K.; writing—original draft preparation, B.S.M., F.R.K.; writing—review and editing, B.S.M., I.S.M., A.E., K.S., F.R.K.; visualization, B.S.M., F.R.K.

Funding: This research received no external funding.

Acknowledgments: The authors thank the volunteers who carried out the beach clean-ups for their hard work and dedication.

Conflicts of Interest: The authors declare no conflict of interest.

References

1. Derraik, J.G.B. The pollution of the marine environment by plastic debris: A review. *Mar. Pollut. Bull.* **2002**, *44*, 842–852. [CrossRef]
2. Do Sul, J.A.I.; Costa, M.F. Marine debris review for Latin America and the Wider Caribbean Region: From the 1970s until now, and where do we go from here? *Mar. Pollut. Bull.* **2007**, *54*, 1087–1104. [CrossRef] [PubMed]
3. Bravo, M.; De losÁngeles Gallardo, M.; Luna-Jorquera, G.; Núñez, P.; Vásquez, N.; Thiel, M. Anthropogenic debris on beaches in the SE Pacific (Chile): Results from a national survey supported by volunteers. *Mar. Pollut. Bull.* **2009**, *58*, 1718–1726. [CrossRef] [PubMed]
4. Araújo, M.C.; Silva-Cavalcanti, J.S.; Costa, M.F. Anthropogenic litter on beaches with different levels of development and use: A snapshot of a coast in Pernambuco (Brazil). *Front. Mar. Sci.* **2018**, *5*, 233. [CrossRef]
5. Ballance, A.; Ryan, P.G.; Turpie, J.K. How much is a clean beach worth? The impact of litter on beach users in the Cape Peninsula, South Africa. *S. Afr. J. Sci.* **2000**, *96*, 210–213.
6. Gregory, M.R.; Andrady, A.L. Plastics in the marine environment. In *Plastics and the Environment*; Andrady, A.L., Ed.; John Wiley and Sons Inc.: Hoboken, NJ, USA, 2003; pp. 379–401.

7. Gregory, M.R. The hazards of persistent marine pollution: Drift plastics and conservation islands. *J. R. Soc. New Zealand* **1991**, *21*, 83–100. [CrossRef]
8. Syberg, K.; Khan, F.R.; Selck, H.; Palmqvist, A.; Banta, G.T.; Daley, J.; Sano, L.; Duhaime, M.B. Microplastics: Addressing ecological risk through lessons learned. *Environ. Toxicol. Chem.* **2015**, *34*, 945–953. [CrossRef]
9. Kordella, S.; Geraga, M.; Papatheodorou, G.; Fakiris, E.; Mitropoulou, I.M. Litter composition and source contribution for 80 beaches in Greece, eastern Mediterranean: A nationwide voluntary clean-up campaign. *Aquat. Ecosyst. Health Manag.* **2013**, *16*, 111–118. [CrossRef]
10. Moore, S.L.; Gregorio, D.; Carreon, M.; Weisberg, S.B.; Leecaster, M.K. Composition and distribution of beach debris in Orange County, California. *Mar. Pollut. Bull.* **2001**, *42*, 241–245. [CrossRef]
11. Nelms, S.E.; Coombes, C.; Foster, L.C.; Galloway, T.S.; Godley, B.J.; Lindeque, P.K.; Witt, M.J. Marine anthropogenic litter on British beaches: A 10-year nationwide assessment using citizen science data. *Sci. Tot. Environ.* **2017**, *579*, 1399–1409. [CrossRef]
12. Ocean Conservancy. *Trassh Travels, International Coastal Cleanup 2010 Report*; The Ocean Conservancy: Washington, DC, USA, 2010.
13. Ocean Conservancy. *The Beach and Beyond, International Coastal Cleanup 2019 Report*; The Ocean Conservancy: Washington, DC, USA, 2019.
14. Rees, G.; Pond, K. Marine litter monitoring programmes – a review of methods with special reference to national surveys. *Mar. Pollut. Bull.* **1995**, *30*, 103–108. [CrossRef]
15. Tudor, D.T.; Williams, A.T. Some threshold levels in beach litter measurement. *Shore Beach* **2001**, *69*, 13–18.
16. Storrier, K.L.; McGlashan, D.J. Development and management of a coastal litter campaign: The voluntary coastal partnership approach. *Mar. Pol.* **2006**, *30*, 189–196. [CrossRef]
17. Syberg, K.; Hansen, S.F.; Christensen, T.B.; Khan, F.R. Risk perception of plastic pollution: Importance of stakeholder involvement and citizen science. In *Freshwater Microplastics the Handbook of Environmental Chemistry*; Wagner, M., Lambert, S., Eds.; Springer: Heidelberg, Germany, 2018; Volume 58, pp. 203–221.
18. Driedger, A.G.; Dürr, H.H.; Mitchell, K.; Van Cappellen, P. Plastic debris in the Laurentian Great Lakes: A review. *J. Gt. Lakes Res.* **2015**, *41*, 9–19. [CrossRef]
19. Bootsma, H.A.; Hecky, R.E. A comparative introduction to the biology and limnology of the African Great Lakes. *J. Gt. Lakes Res.* **2003**, *29*, 3–18. [CrossRef]
20. Lusher, A.L.; McHugh, M.; Thompson, R.C. Occurrence of microplastics in the gastrointestinal tract of pelagic and demersal fish from the English Channel. *Mar. Pollut. Bull.* **2013**, *67*, 94–99. [CrossRef] [PubMed]
21. Sanchez, W.; Bender, C.; Porcher, J.M. Wild gudgeons (Gobiogobio) from French rivers are contaminated by microplastics: Preliminary study and first evidence. *Environ. Res.* **2014**, *128*, 98–100. [CrossRef]
22. Güven, O.; Gökdağ, K.; Jovanović, B.; Kıdeyş, A.E. Microplastic litter composition of the Turkish territorial waters of the Mediterranean Sea, and its occurrence in the gastrointestinal tract of fish. *Environ. Pollut.* **2017**, *223*, 286–294. [CrossRef]
23. Calderon, E.A.; Hansen, P.; Rodríguez, A.; Blettler, M.C.; Syberg, K.; Khan, F.R. Microplastics in the digestive tracts of four fish species from the Ciénaga Grande de Santa Marta Estuary in Colombia. *Water Air Soil Pollut.* **2019**, *230*, 257. [CrossRef]
24. Biginagwa, F.; Mayoma, B.; Shashoua, Y.; Syberg, K.; Khan, F.R. First evidence of microplastics in the African Great Lakes: Recovery from Lake Victoria Nile perch and Nile tilapia. *J. Gt. Lakes Res.* **2016**, *42*, 1146–1149. [CrossRef]
25. Khan, F.R.; Mayoma, B.S.; Biginagwa, F.J.; Syberg, K. Microplastics in Inland African Waters: Presence, Sources, and Fate. In *Freshwater Microplastics the Handbook of Environmental Chemistry*; Wagner, M., Lambert, S., Eds.; Springer: Heidelberg, Germany, 2018; Volume 58, pp. 101–124.
26. Ngupula, G.W.; Kayanda, R.J.; Mashafi, C.A. Abundance, composition and distribution of solid wastes in the Tanzanian waters of Lake Victoria. *Afr. J. Aquat. Sci.* **2014**, *39*, 229–232. [CrossRef]
27. Lewis, J. *Application of Environment Agency's Aesthetic Assessment Protocol (Beach Survey).R, D Technical Summary E1-117/TR*; Environment Agency: Bristol, UK, 2002; p. 3.
28. Silva-Iniguez, L.; Fisher, D.W. Quantification and classification of marine litter on the municipal beach of Ensenada, Baja California, Mexico. *Mar. Pollut. Bull.* **2003**, *46*, 132–138. [CrossRef]
29. Fujieda, S.; Sasaki, K. Stranded debris of foamed plastic on the coast of ETA Island and Kurahashi Island in Hiroshima Bay. *Nippon. Suisan Gakkaishi* **2005**, *71*, 755–761. [CrossRef]

30. Browne, M.A.; Galloway, T.S.; Thompson, R.C. Spatial patterns of plastic debris along estuarine shorelines. *Environ. Sci. Technol.* **2010**, *44*, 3404–3409. [CrossRef] [PubMed]
31. Free, C.M.; Jensen, O.P.; Mason, S.A.; Eriksen, M.; Williamson, N.J.; Boldgiv, B. High-levels of microplastic pollution in a large, remote, mountain lake. *Mar. Pollut. Bull.* **2014**, *85*, 156–163. [CrossRef]
32. Njeru, J. The urban political ecology of plastic bag waste problem in Nairobi, Kenya. *Geoforum* **2006**, *37*, 1046–1058. [CrossRef]
33. Clapp, J.; Swanston, L. Doing away with plastic shopping bags: International patterns of norm emergence and policy implementation. *Environ. Polit.* **2009**, *18*, 315–332. [CrossRef]
34. Dikgang, J.; Leiman, A.; Visser, M. Resources, conservation and recycling analysis of the plastic-bag levy in South Africa. *Resour. Conserv. Recycl.* **2012**, *66*, 59–65. [CrossRef]
35. Tanzania Bans Tourists From Bringing Plastic Bags Into the Country. Available online: https://www.globalcitizen.org/en/content/tanzania-plastic-bag-ban-travelers/ (accessed on 22 November 2019).
36. NOAA Marine Debris Tracker. Available online: https://marinedebris.noaa.gov/partnerships/marine-debris-tracker (accessed on 22 November 2019).
37. Marine LitterWatch in a Nutshell. Available online: https://www.eea.europa.eu/themes/water/europes-seas-and-coasts/assessments/marine-litterwatch/at-a-glance/marine-litterwatch-in-a-nutshell (accessed on 24 November 2019).
38. Ruiz-Mallén, I.; Riboli-Sasco, L.; Ribrault, C.; Heras, M.; Laguna, D.; Perié, L. Citizen science: Toward transformative learning. *Sci. Commun.* **2016**, *38*, 523–534. [CrossRef]
39. Beat Plastic Pollution. Available online: www.unenvironment.org/interactive/beat-plastic-pollution (accessed on 22 November 2019).
40. Adebiyi-Abiola, B.; Assefa, S.; Sheikh, K.; García, J.M. Cleaning up plastic pollution in Africa. *Science* **2019**, *365*, 1249–1251. [CrossRef]

Article

'The Plastic Nile': First Evidence of Microplastic Contamination in Fish from the Nile River (Cairo, Egypt)

Farhan R. Khan [1,*], Yvonne Shashoua [2], Alex Crawford [3], Anna Drury [3], Kevin Sheppard [3], Kenneth Stewart [3] and Toby Sculthorp [3]

1 Department of Science and Environment, Roskilde University, Universitetsvej 1, P.O. Box 260, DK-4000 Roskilde, Denmark
2 Environmental Archaeology and Materials Science, National Museum of Denmark, IC Modewegsvej Brede, DK-2800 Kongens Lyngby, Denmark; Yvonne.Shashoua@natmus.dk
3 Sky News International, Grant Way, Islwwroth, Middlesex TW7 5QD, UK; Alex.Crawford@sky.uk (A.C.); annamc25@hotmail.com (A.D.); Kevin.Sheppard@sky.uk (K.S.); Kenneth.Stewart@sky.uk (K.S.); Toby.Sculthorp@sky.uk (T.S.)
* Correspondence: frkhan@ruc.dk; Tel.: +45-5019-0429

Received: 17 September 2019; Accepted: 12 November 2019; Published: 25 March 2020

Abstract: The presence of microplastics (MPs) in the world's longest river, the Nile River, has yet to be reported. This small-scale study aimed to provide the first information about MPs in the Nile River by sampling the digestive tracts of two fish species, the Nile tilapia (*Oreochromis niloticus*, $n = 29$) and catfish (*Bagrus bayad*, $n = 14$). Fish were purchased from local sellers in Cairo, and then their gastrointestinal tracts were dissected and examined for MPs. Over 75% of the fish sampled contained MPs in their digestive tract (MP prevalence of 75.9% and 78.6% for Nile tilapia and catfish, respectively). The most abundant MP type was fibers (65%), the next most abundant type was films (26.5%), and the remaining MPs were fragments. Polyethylene (PE), polyethylene terephthalate (PET) and polypropylene (PP) were all non-destructively identified by attenuated total reflectance Fourier transform infrared spectroscopy. A comparison with similar studies from marine and freshwater environments shows that this high level of MP ingestion is rarely found and that fish sampled from the Nile River in Cairo are potentially among the most in danger of consuming MPs worldwide. Further research needs to be conducted, but, in order to mitigate microplastic pollution in the Nile River, we must act now.

Keywords: microplastics; freshwater; Africa; ingestion; Nile tilapia (*Oreochromis niloticus*); catfish (*Bagrus Bajad*); fibers; ATR-FTIR spectroscopy

1. Introduction

Microplastic pollution (MPs, defined as <5 mm in size) has been found in all compartments of the aquatic environment—water, sediment and the animals that inhabit them. Microplastics have been found in the depths of the ocean [1,2] and in remote mountain locations [3,4]. However, some significant knowledge gaps relating to the presence and abundance of MPs across our planet remain, both in terms of aquatic habitat and geographical location. Most of the scientific research into MP pollution still focuses on the marine environment, and less is known about MPs in freshwaters [5]. MP research is largely concentrated in Europe and North America, and is under-represented in Africa, Asia and South America [5]. Despite the fact that the primary focus of MP research has been in the marine environment, there has been increasing interest in determining the presence MPs in river systems. Countless people depend on rivers for a variety of vital functions, and, furthermore, rivers have been shown as major pathways for the transport of substantial amounts of plastic debris and MPs from land-based sources to the oceans [6,7]. Thus, in recent years, MPs have been documented in the water, sediment and biota of the some of the world's major river systems; the Thames (UK) [8,9], the Seine (France) [10], the Rhine (Germany) [11] and the Danube (Austria) [12] in Europe; the Amazon in South America [13,14]; the Yangzte in Asia (China) [15,16]; and the St. Lawrence in North America [17]. One notable river not present on this list, arguably the most famous river of all, is the Nile River. The aim of this small-scale study was to rectify this knowledge gap and provide the first information about MPs in the Nile River.

The Nile River is the longest river in the world at 6693 km [18], and it is the only permanent river to cross the Sahara Desert [19]. The two major tributaries of the Nile are the White Nile and the Blue Nile (Figure 1). While the latter's origin in Lake Tana in Ethiopia is well known, the origin of the longer White Nile is debatable, but it is considered to be a tributary of the Kagera River which flows into Lake Victoria [19]. The White Nile tributary then leaves the African Great Lake at the Ugandan city of Jinja, flows northwards through South Sudan and its capital Juba, and then flows on through Sudan where the White and Blue Nile tributaries join at the Sudanese capital Khartoum, which has an estimated population of close to 6 million people [20]. The Nile continues north into Egypt, passing through Aswan before flowing through Cairo, which has a population estimated to be approximately 20 million people by 2020 [20]. The Nile River drains into Mediterranean Sea via the Nile Delta (Figure 1). Whilst the Nile flows through numerous countries, it has always been inextricably linked to Egypt and has been described as the "donor of life to Egypt" [21]. The role of the Nile in establishing the Ancient Egyptian civilizations cannot be understated, and the historic dependence on the Nile continues today through agriculture, transport, fishing and tourism [19]. However, the adverse impacts of plastic and MPs could be a pervasive threat that have not yet been researched. Here, we investigate the presence of MPs by looking within the gastrointestinal (digestive) tracts of two fish species found in the Nile River, the Nile tilapia (*Oreochromis niloticus*) and catfish (*Bagrus Bajad*).

Sampling the digestive tracts of resident fish populations has become a recognized approach by which to assess the extent of MP pollution in the environment. Examples of this type of study have found MPs in the gastrointestinal tracts of fish from the English Channel [22], French freshwater systems [23], the Mediterranean coast of Turkey [24], the Amazon River estuary [13] and Lake Victoria [25]. This last study was first to document the presence of MPs in African freshwaters and was conducted in the largest of Africa's Great Lakes. Nile perch (*Lates niloticus*) and Nile tilapia were selected for their economic and ecological importance and were purchased from the local market in the Mwanza region of Tanzania. Gastrointestinal tracts were dissected and then digested in a strong alkaline solution to isolate MPs. In total, suspected plastics were recovered from the gastrointestinal tracts of 11 perch (55%) and seven tilapia (35%), and they were confirmed in 20% of each fish species (i.e., four individuals) by attenuated total reflectance Fourier transform infrared (ATR-FTIR) spectroscopy, the definitive analytical technique for identifying the chemical composition of plastic polymers. A variety of polymers were recovered from the fish including polyethylene, polyurethane and polyester, which are found in packaging, clothing, and food and drink containers. The likely sources of these plastics are considered to be human activities linked to fishing and tourism, as well as urban waste [25]. Though limited in the number of fish used, the Lake Victoria study exemplified how a small-scale investigation could be conducted and provide an early indications of MP pollution [26].

The present study was conducted with fish caught in the center of Cairo and used the same methods as in Lake Victoria. Nile tilapia were once again utilized, and the second species used was the catfish, *Bagrus Bayad*. Nile tilapia typically inhabit shallow waters and are omnivorous with a diet consisting of plankton and smaller fish [27], whereas the much larger catfish are exclusively piscivorous, living and feeding near the bottom of the water column [28]. The presence of MPs in these two species of differing niches and feeding habitats could provide additional information by which to better understand the fate of MPs in freshwater rivers.

This study was conceived as part of the documentary 'The Plastic Nile' produced by Sky News International. It was conducted within a relatively short period of time (five days) and under a degree of confidentiality that is not usual for scientific investigations. The reason for this is that criticism of the Nile is often negatively received by the authorities, as evidenced by the case of Egyptian singer Sherine Abdel Wahab, who was sentenced to six months imprisonment for commenting on the cleanliness of the Nile River [29]. Thus, the work was performed in a laboratory within the greater Cairo area, but no further details are provided as to not identify our hosts. Moreover, they have requested not to be added as co-authors to this work, but we greatly acknowledge and appreciate their collaboration, without which this work would not have been possible.

Figure 1. Path of the Nile River starting at Lake Tana (Blue Nile) and Lake Victoria (White Nile), respectively, before meeting at Khartoum and flowing toward the Mediterranean Sea via Cairo, the area of the present study (**A**). Fish were caught around Dahab Island in the Nile River at Cairo and purchased from markets located along the Nile Corniche on the east bank of the Nile opposite Dahab Island (**B**). Examples of plastic litter found close to Dahab Island and along the Nile (**C**,**D**). (Map source: Google Earth (2019) with the path of the Nile River overlaid from shapefiles [30]).

2. Materials and Methods

2.1. Study Area and Fish

In December 2018, twenty-nine Nile tilapia (*Oreochromis niloticus*) and fourteen catfish (*Bagrus bayad*) were purchased from local sellers situated along the Nile Corniche in Cairo, where fish are caught and sold daily. The fish were caught in the vicinity of Dahab Island, which is located in the heart of Cairo and close to The Great Pyramid of Giza (Figure 1). Fish were purchased whole without the prior removal of their gastrointestinal tracts (i.e., they were not gutted) and transported to the laboratory, where they were promptly refrigerated at 4 °C. The gastrointestinal tracts of each fish were removed within 24 h of arrival in the laboratory. Prior to dissection, all fish were measured and weighed. Nile tilapia (n = 29) had an average weight and length of 129.7 ± 59.0 g and 17.7 ± 2.5 cm, respectively, and the catfish (n = 14) were considerably larger at 1496.4 ± 849.2 g and 56.5 ± 18.9 cm. The catfish weight exceeded the limit of the digital laboratory balance and were therefore weighed on an analog kitchen scale. The heaviest catfish was just over 3000 g. The length and weight of each individual fish used in this study can be found in Table S1 (Supplementary Information).

2.2. Tissue Digestion and MP Extraction

The methods for dissection and digestion, as well as the subsequent isolation of MPs, were conducted as previously described by Biginagwa et al. (2015) [25]. For each fish, the entire gastrointestinal tract from buccal cavity to anus was dissected following a longitudinal incision of the abdomen. All efforts were made to eliminate sample contamination with the thorough cleaning of dissection utensils with 70% ethanol and lint-free paper between dissections. A preliminary examination was made of each gastrointestinal tract, and, in the case of catfish, undigested smaller fish were removed. Dissected tissues were placed in 250 or 500 mL conical flasks, to which 10 M NaOH was added in a 5:1 (*w*/*v*) ratio. The NaOH digestion (60 °C for 24 h) was used to isolate plastic litter from the organic tissue. This method, which involves a strong basic solution, has been shown to digest organic matter with an efficacy of >90% [15,31] whilst importantly having negligible impact on the plastics, especially when compared to strong acid digestion, which can discolor or degrade plastics. Post-digestion, the plastics and a minimal amount of partially digested tissue were rinsed from the NaOH through 250 μm mesh stainless steel sieves under running water and then placed on filter paper (Whatman® Grade 540, Hardened Ashless Filter Paper, 90 mm diameter, GE Healthcare Life Sciences, UK) to dry under a fume hood. The dried samples were then tightly wrapped within the filter papers, sealed in 50 mL falcon tunes, and transported to the laboratory at Roskilde University (Denmark).

In the laboratory, each sample was examined under a light dissection microscope (×40). MPs were initially visually identified due to their possession of unnatural coloration (such as bright blue) and/or unnatural shapes (such as fragments with sharp edges) [32,33]. A secondary visual inspection was made in which each suspected item was required to possess the following criteria as described by Nor and Obbard (2014) [34] and Horton et al. (2018) [35]: (1) no visible cellular or organic structures, (2) unsegmented, (3) fibers of homogenous width, (4) the appearance of homogenous material, (5) fibers that remained intact if pulled with tweezers, and (6) flexible, but not brittle. These two steps were used to document the presence of MPs. MPs found per fish were enumerated and categorized as fibers, fragments, films, pellets, foams or beads, according to Tanaka and Takada (2016) [36]. Pictures of the suspected MPs were taken with an Olympus uc90 digital camera mounted on an Olympus SZ61 microscope. A subsample of MPs was analyzed by Fourier transform infrared spectroscopy to verify the visual identification based on its chemical structure.

2.3. Fourier Transform Infrared Spectroscopy

The chemical composition of a representative sample of MPs was non-destructively identified by attenuated total reflectance Fourier transform infrared (ATR-FTIR) spectroscopy. ATR-FTIR has become a standard analytical technique for identifying the chemical composition of samples larger than 0.5 mm. However, owing to this size constraint, the representative sample analyzed was made from those plastic particles that were at least 0.5 mm in one dimension. In total, 10% of the collected MPs were identified by ATR-FTIR. Scans were run at a resolution of 2 cm^{-1} between 4000 and 650 cm^{-1} on a Bruker Alpha-p FTIR spectrometer (Bruker, Billirica, MA, USA) fitted with a diamond single-bounce internal reflectance element. Spectra were compared with reference standards run on the same instrument and processed using Opus software supplied by Bruker. To make a positive polymer identification, we used the approach advocated by Frias et al. (2016) [37] and the European Union expert group on marine litter (Subgroup on Marine litter (TSG-ML)) to only accept matches of >70% similarity to the reference library samples [38].

2.4. *Quality Assurance Measures*

Neoprene gloves and cotton lab coats were worn during the dissection, filtering, and microscope processes in order to avoid contamination. All glassware used was washed and rinsed with distilled water. During the process to isolated MPs, six background samples (petri dishes containing filter papers) were placed in the lab space. This included around the dissection area and under the fume hood whilst the samples dried to account for any airborne MPs. The references were transported along with the samples back to Denmark. All background samples were systematically examined under light dissection microscope all with the other samples. Only one black fiber was found in the reference petri dishes.

2.5. *Analysis*

The number of individuals containing MPs from each species was expressed as the frequency of occurrence of MPs (FO%), as follows: FO% = (Ni/N) × 100, where Ni is the number of digestive tracts that contained MPs and N is the total number of gastrointestinal tracts examined [13]. Differences in FO% and the average number of items found per individual between Nile tilapia and catfish were assessed by an unpaired *t*-test (SPSS version 22 (SPSS statistics for windows, SPSS Inc., Chicago, IL, USA)).

3. Results and Discussion

3.1. *Abundance of MPs in Fish Gastrointestinal Tracts*

Over 75% of the 43 fish sampled in this study contained MPs in their gastrointestinal tract (33 out of 43, FO = 76.7%). From these 33 fish, a total of 211 items of plastic were recovered. The highest number of MPs recovered from a single fish was 20 individual items, which were found in a Nile tilapia (sample #15, Table S1). The two species showed similar levels of MP prevalence, with FO values of 75.9% and 78.6% for Nile tilapia and catfish, respectively. Of the 22 out of 29 Nile tilapia that contained MPs in their gastrointestinal tract, 164 MPs were recovered. Thus, each tilapia that contained MPs contained, on average, 7.5 ± 4.9 items. Forty-seven MPs were recovered from the 11 out of 14 catfish with MPs in their digestive tracts, with each individual containing MPs having an average burden of 4.7 ± 1.7 items (Table 1). This difference between 7.5 ± 4.9 and 4.7 ± 1.7 items per fish was significant ($p = 0.046$; unpaired *t*-test), thus suggesting that Nile tilapia either consume more MPs than catfish or better retain them within their digestive system. As Nile tilapia inhabit shallower waters closer to the surface, they may be more likely to come into contact with MPs than the bottom-dwelling catfish. Moreover, the omnivorous diet of the tilapia which contains plankton may mean that it is more likely to mistake plastic items for food rather than the strictly piscivorous catfish. Any species differences should be treated with caution due to the limited numbers of fish involved in this study. However, a comparison of feeding type did find that omnivorous fish with their diverse diets and higher feeding rates were more likely to contain MP fibers than either herbivorous or carnivorous species [39].

Table 1. Summary of data for Nile tilapia and catfish. The number of individual fish analyzed (n), average weights, and average lengths (± standard deviation) are provided. The frequency of occurrence (in %) describes the number of fish with microplastics (MPs) present in their gastrointestinal tracts. The total numbers of MPs found per species and average per individual are given alongside a breakdown of the different MP types found (in %). The full dataset of MPs found in each individual fish can be found in Supplemental Information (Table S1).

Common Name	Species Name	*n*	Weight (g)	Length (cm)	% FO	Total Number of MPs Found	Mean MPs Per Individual Found with MPs	Type (%)		
								Fibers	Films	Frag.
Nile tilapia	*Oreochromis niloticus*	29	129.7 ± 59.0	17.7 ± 2.5	75.9	164	7.5 ± 4.9	65.3	25.6	8.5
Catfish	*Bagrus Bajad*	14	1496.4 ± 849.2	56.5 ± 18.9	78.6	47	4.7 ± 1.7	61.7	29.8	8.5

The frequency of occurrence in fish sampled from the Nile River at Cairo appears to generally higher than those reported from other locations. Fish sampled from the marine environment have wide ranging values; at the low end of the scale, only 2.6% of fish sampled from the North Sea [40] and only 5.5% of fish in the North and Baltic Seas [41] contained MPs. Elsewhere, the prevalence of MPs in different fish species has been determined to be 19.8% of fish from the Portuguese coast [42], 37% from the English Channel [22], and 68% from the Balearic Islands [43]. The Mediterranean Sea, into which the Nile River flows, is perhaps the most valid marine comparison to be made. An investigation of MP prevalence in fish from the Turkish waters of the Mediterranean Sea found that 41% of sampled fish representing 28 species contained MPs [24]. In Spanish and Mediterranean coastal waters, 17.5% of examined fish contained MPs in their digestive tracts, with the highest occurrence in red mullet (18.8%) [44]. The larger pelagic fish of the Mediterranean—swordfish, Bluefin tuna and albacore—had occurrences of 12.5%, 32.4% and 12.9%, respectively [45].

In comparing studies, it is important to make relevant comparisons. For the Nile River, the two most relevant comparisons are studies in which fish were sampled from freshwater rivers or conducted in locations with similar sized urban populations. For each of these parameters, there are only a handful of relevant findings to compare against. Perhaps the only comparable river to the Nile in the world is the Amazon, and the two have both been described as the world's longest river. However, the MP prevalence in fish from the Amazon estuary and northern coast of Brazil is markedly less than those from the Nile, as MPs have only been found in 26 out of 189 examined gastrointestinal tracts (13.8%) across 14 different species [13]. Fish sampled from French and Belgian freshwater river systems have been shown to have an overall FO of only 12% and 9%, respectively [23,46] and, of the fish sampled from the River Thames (UK), 33% contained MPs [35]. However, whist the Thames is considered to be a major river and the city of London has a sizeable population, the sampling sites for this study were situated outside of the largest urban areas. Research conducted in Tokyo Bay (Japan) with a population in the bay's drainage area of 29 million people found that 77% of Japanese anchovies contained MPs in their digestive tracts [36]. Very few studies have reported higher occurrences, but in the Río de la Plata [47] and Bahía Blanca [48] estuaries in Argentina, MPs have been found in 100% of the sampled fish. Notably, the study by Pazos et al. (2017) [47] reported an average number of MPs per fish as 18.5 ± 18.9, whilst more typical numbers were found to be between 2 and 10. Our results predominantly fit in this range.

Comparing a wide range of aquatic habitats and fish species, it appears that the frequencies of occurrence for MPs in Nile tilapia and catfish sampled from the urban waters of Cairo are amongst the higher levels reported in the scientific literature.

3.2. Characteristics and Identification of Polymer Types

Fibers were the most commonly found MP type, accounting for 65.3% and 61.7% of items found in the gastrointestinal tracts of Nile tilapia and catfish, respectively. Films were the second most abundant, with 25.6% (Nile tilapia) and 29.8% (catfish), and fragments made up 8.5% of the recovered items in both species (Table 1). Pellets, foams and beads were not found. Black and red colored MPs were most abundant for fibers and films (black > red > blue > green > other > transparent for fibers and black > red > transparent > green > blue for films). Fragments were predominantly blue (blue > black > transparent). The full dataset for MP number, type and color per individual can be found in Table S1 (Supplementary Information). A selection of the MPs recovered from the digestive tracts of the sampled fish can be found in Figure 2.

The dominance of fibers as the most abundant MP type is well founded in the literature. Fibers constituted the greatest proportion of MPs in fish from the Río de la Plata estuary (96%) [47] and in the North Pacific Gyre (94%) [49]. The studies by Neves et al. (2015) [42], Bellas et al. (2016) [44] and Arias et al. (2019) [48] also reported greater numbers of fibers compared to other MP types within piscine digestive tracts. The reason for fibers being so common has been attributed to their diverse origin. Fibers may result from the degradation of clothing items, furniture and fishing gear. Washing a single item of synthetic clothing may release approximately 2000 fibers [50].

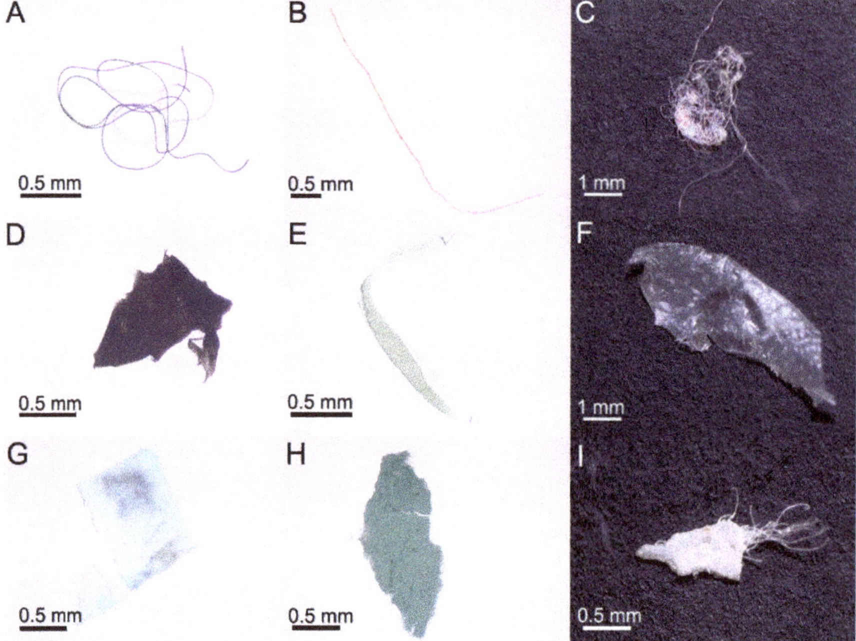

Figure 2. Examples of MPs found within the gastrointestinal tracts of Nile Tilapia and Catfish: fibers (blue polyethylene (PE) (**A**), Red polyethylene terephthalate (PET) (**B**), and PET fiber knot (**C**)), films (black (**D**), green (**E**) and transparent (**F**) films were all identified as PE), and fragments (light blue polypropylene (PP) (**G**), blue-green PP (H) and white PET (**I**)).

ATR-FTIR spectroscopy was used to verify the visual selection of MPs. All of the analyzed MPs (approximately 10% of the 211 recovered MPs) were plastic polymers. Polyethylene (PE), polyethylene terephthalate (PET), and polypropylene (PP) were all found (Figures 2 and 3) in keeping with other studies. These polymers are used in a multitude of consumer products from clothing, to packaging, to plastic bags, as well as in fishing nets and ropes. Given the diverse usage of commonly found polymers, it is not possible to directly attribute the presence of MPs within fish digestive tracts to any specific source.

ATR-FTIR also revealed that a number of the MPs had undergone a degree of weathering and chemical oxidation (Figure 3). For example, the spectrum of the blue PP fragment had developed additional broad peaks at 1620–1750 cm^{-1} and 1100 cm^{-1} that may be attributed to degradation compared to the pristine PP reference material. MPs in African freshwaters, as in other equatorial locations, are likely to degrade faster than in more temperate conditions because reactions such as photolysis, thermo-oxidation and photo-oxidation are accelerated by intensive UV light [51]. The implications of weathering are not well understood, but environmentally-aged or degraded particles may have the rate at which they sink though the water column altered, which in turn may effect which organisms they meet. Furthermore, weathered MPs maybe more susceptible to biofouling (the attachment of microorganisms to the surface of the MP) or the adsorption of other environmental pollutants such as hydrophobic, organic contaminants or trace metals [52,53].

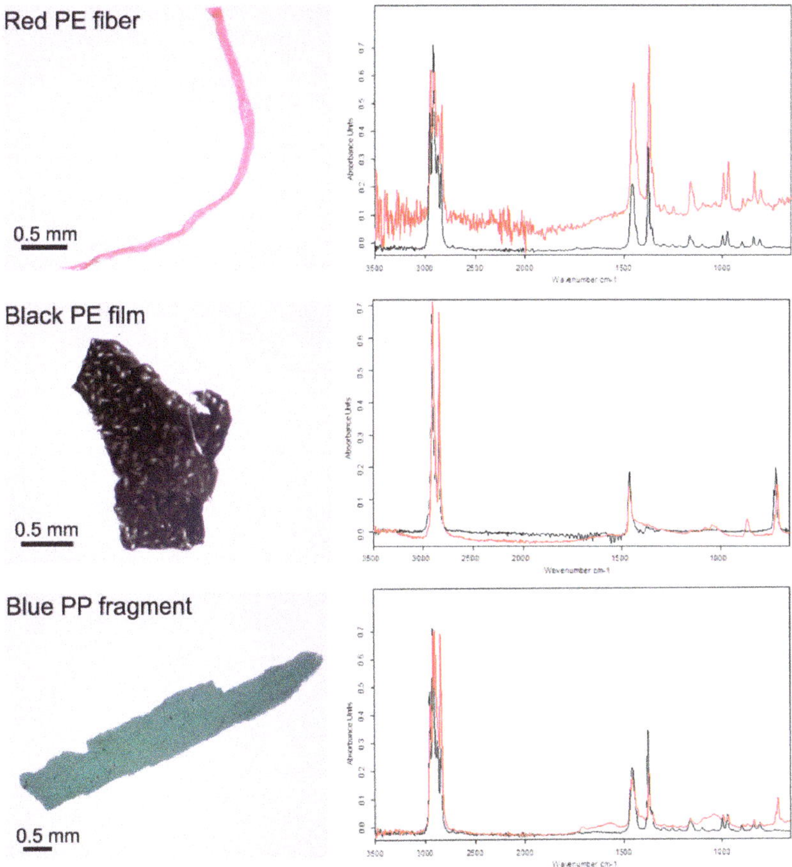

Figure 3. Fourier-transform infrared (FTIR) spectra of representative samples identified as polyethylene (PE) (both red fiber and black film) and polypropylene (PP) (blue-green fragment). Sample spectra are shown as red and compared to known reference samples in black. Samples exhibited weathering and oxidation, as shown by the presence of additional broad peaks, e.g., at wavelengths of 1620–1750 cm^{-1} and 1100 cm^{-1} in the spectra of the blue PP fragment.

3.3. Future Considerations

Our study provides the first assessment of MP pollution in the Nile River. Over 75% of the sampled fish contained at least one item of plastics in their gastrointestinal tract, and five of the Nile tilapia contained 10 MPs or more (Table S1, Supplemental Information). Comparisons across studies from marine and freshwater environments show that this level of MP ingestion is rarely found and that fish sampled from the Nile River in Cairo are potentially among the most in danger of consuming MPs on the planet. However, ours was only a small-scale study, and, although this preliminary data present a worrying scenario, we likely underestimate the true extent of plastic ingestion by Nile tilapia and catfish. Here, we were constrained by time, fish availability and the size-selected MPs of above 250 μm. Thus, the MPs below this size were not captured, and, based on previous research, this is likely to be a sizeable fraction of environmentally-present MPs [54]. Moreover, smaller-sized plastic items, such as those found in the nano-size range (less than 1 μm), have the potential to cross intestinal barriers and enter fish tissue, which could, in turn, allow MPs to enter the human food chain [55]. Since most fish, including those used in this study, are eaten following the removal of the digestive tract, the presence of larger MPs is less concerning in terms human consumption, but the presence of MPs in fish digestive tracts is in itself a disturbing phenomenon that has been reported from many locations.

As well as being the first to describe the presence of MPs in fish from the Nile, this study is only the second to describe MPs within fish from African freshwaters—the first having been conducted in Lake Victoria [25]. Other research has quantified and characterized the presence of MPs in African freshwaters, such as within gastropods from the Osun River system in Nigeria [56] and within the sediments of the lagoon of Bizerte in Tunisia [57]. However, such studies are scarce, and there is a pressing need for more information regarding the prevalence of MPs in Africa's inland freshwaters [26].

The present study, as well those few others conducted in Africa, present only a 'snapshot in time' of MP pollution in African waters. Based on our preliminary findings from Cairo, but also with the broader perspective encompassing African freshwaters, we suggest that establishing a complete picture of MP pollution along the Nile River should be considered a research priority. This may be through a continuous environmental monitoring program for MP pollution that encompasses water, sediment, and biota, as well as sampling at a number of sites—particularly those with a dense urban population.

The confirmation of MPs in the Nile River is only the first, albeit necessary, step of starting to understand and stop plastic pollution in the world's longest river. Longer-term research needs to be conducted that also encompasses the impacts of MP pollution on fish populations and its potential transfer to surrounding human populations. Understanding the sources and fate of MPs that enter the Nile is key to mitigating its impacts, but this understanding requires the collaboration of numerous interested stakeholders. The Nile River is part of the human story, and the advent of the 'plastic age' may result in impacts on this ancient river that are not yet possible to predict. It may already be too late to prevent such outcomes, but if we are to succeed, we must act without delay.

Supplementary Materials: The following are available online at http://www.mdpi.com/2305-6304/8/2/22/s1, Table S1: Full dataset of microplastic number, type and color found in each individual Nile tilapia (A) and catfish (B).

Author Contributions: Conceptualization, A.C., A.D., K.S. (Kevin Sheppard), K.S. (Kenneth Stewart), T.S., F.R.K.; investigation, F.R.K.; methodology, F.R.K.; validation, F.R.K, Y.S.; resources, A.C., A.D., K.S. (Kevin Sheppard), K.S. (Kenneth Stewart), T.S., F.R.K.; writing—original draft preparation, F.R.K.; writing—review and editing, F.R.K., Y.S., A.C., A.D., K.S. (Kevin Sheppard), K.S. (Kenneth Stewart), T.S.; visualization, F.R.K.; funding acquisition, A.C., T.S.

Funding: This research was funded by Sky News International (no grant number).

Acknowledgments: The authors reiterate their gratitude to the unnamed collaborators without which this work would not have been possible.

Conflicts of Interest: The present study was commissioned and conceived as part of the documentary 'The Plastic Nile' produced by Sky News International. The authors from Sky News were involved in the conceptualization of the study and the organizational logistics in Cairo. However the scientific research was conducted independently of the documentary. FRK was remunerated as a Scientific Advisor for the documentary.

References

1. Woodall, L.C.; Sanchez-Vidal, A.; Canals, M.; Paterson, G.L.J.; Coppock, R.; Sleight, V.; Calafat, A.; Rogers, A.D.; Narayanaswamy, B.E.; Thompson, R.C. The deep sea is a major sink for microplastic debris. *R. Soc. Open Sci.* **2014**, *1*, 140317. [CrossRef] [PubMed]
2. Peng, X.; Chen, M.; Chen, S.; Dasgupta, S.; Xu, H.; Ta, K.; Bai, S. Microplastics contaminate the deepest part of the world's ocean. *Geochem. Perspect. Lett.* **2018**, *9*. [CrossRef]
3. Free, C.M.; Jensen, O.P.; Mason, S.A.; Eriksen, M.; Williamson, N.J.; Boldgiv, B. High levels of microplastic pollution in a large, remote, mountain lake. *Mar. Pollut. Bull.* **2014**, *85*, 156–163. [CrossRef] [PubMed]
4. Zhang, K.; Su, J.; Xiong, X.; Wu, X.; Wu, C.; Liu, J. Microplastic pollution of lakeshore sediments from remote lakes in Tibet plateau, China. *Environ. Pollut.* **2016**, *219*, 450–455. [CrossRef]
5. Blettler, M.C.M.; Abrial, E.; Khan, F.R.; Sivri, N.; Espinola, L.A. Freshwater plastic pollution: Recognizing research biases and identifying knowledge gaps. *Water Res.* **2018**, *143*, 416–424. [CrossRef] [PubMed]

6. Schmidt, C.; Krauth, T.; Wagner, S. Export of plastic debris by rivers into the sea. *Environ. Sci. Technol.* **2017**, *51*, 12246–12253. [CrossRef]
7. Lebreton, L.C.; Van der Zwet, J.; Damsteeg, J.W.; Slat, B.; Andrady, A.; Reisser, J. River plastic emissions to the world's oceans. *Nat. Commun.* **2017**, *8*, 15611. [CrossRef]
8. Morritt, D.; Stefanoudis, P.V.; Pearce, D.; Crimmen, O.A.; Clark, P.F. Plastic in the Thames: A river runs through it. *Mar. Pollut. Bull.* **2014**, *78*, 196–200. [CrossRef]
9. Horton, A.A.; Svendsen, C.; Williams, R.J.; Spurgeon, D.J.; Lahive, E. Large microplastic particles in sediments of tributaries of the River Thames, UK—Abundance, sources and methods for effective quantification. *Mar. Pollut. Bull.* **2017**, *114*, 218–226. [CrossRef]
10. Dris, R.; Gasperi, J.; Rocher, V.; Saad, M.; Renault, N.; Tassin, B. Microplastic contamination in an urban area: A case study in Greater Paris. *Environ. Chem.* **2015**, *12*, 592–599. [CrossRef]
11. Mani, T.; Hauk, A.; Walter, U.; Burkhardt-Holm, P. Microplastics profile along the Rhine River. *Sci. Rep.* **2015**, *5*, 17988. [CrossRef] [PubMed]
12. Lechner, A.; Ramler, D. The discharge of certain amounts of industrial microplastic from a production plant into the River Danube is permitted by the Austrian legislation. *Environ. Pollut.* **2015**, *200*, 159–160. [CrossRef] [PubMed]
13. Schmid, K.; Winemiller, K.O.; Chelazzi, D.; Cincinelli, A.; Dei, L.; Giarrizzo, T. First evidence of microplastic ingestion byfishes from the Amazon River estuary. *Mar. Pollut. Bull.* **2018**, *133*, 814–821.
14. Andrade, M.C.; Winemiller, K.O.; Barbosa, P.S.; Fortunati, A.; Chelazzi, D.; Cincinelli, A.; Giarrizzo, T. First account of plastic pollution impacting freshwater fishes in the Amazon: Ingestion of plastic debris by piranhas and other serrasalmids with diverse feeding habits. *Environ. Pollut.* **2019**, *244*, 766–773. [CrossRef] [PubMed]
15. Zhao, S.; Zhu, L.; Wang, T.; Li, D. Suspended microplastics in the surface water of the Yangtze Estuary System, China: First observations on occurrence, distribution. *Mar. Pollut. Bull.* **2014**, *86*, 562–568. [CrossRef] [PubMed]
16. Zhang, K.; Gong, W.; Lv, J.; Xiong, X.; Wu, C. Accumulation of floating microplastics behind the Three Gorges Dam. *Environ. Pollut.* **2015**, *204*, 117–123. [CrossRef] [PubMed]
17. Castañeda, R.A.; Avlijas, S.; Simard, M.A.; Ricciardi, A. Microplastic pollution in St. Lawrence river sediments. *Can. J. Fish. Aquat. Sci.* **2014**, *71*, 1767–1771. [CrossRef]
18. World Atlas. Available online: https://www.worldatlas.com/articles/which-are-the-longest-rivers-in-the-world.html (accessed on 28 August 2019).
19. Dumont, H.J. *The Nile: Origin, Environments, Limnology and Human Use*; Springer: Dordrecht, The Netherlands, 2009.
20. United Nations, Department of Economic and Social Affairs, Population Division. 2019. Available online: https://esa.un.org/unpd/wup/ (accessed on 28 August 2019).
21. Osman, A.G.M.; Kloas, W. Water Quality and Heavy Metal Monitoring in Water, Sediments, and Tissues of the African Catfish Clarias gariepinus (Burchell, 1822) from the River Nile, Egypt. *J. Environ. Prot.* **2010**, *1*, 389–400. [CrossRef]
22. Lusher, A.L.; McHugh, M.; Thompson, R.C. Occurrence of microplastics in the gastrointestinal tract of pelagic and demersal fish from the English Channel. *Mar. Pollut. Bull.* **2013**, *67*, 94–99. [CrossRef]
23. Sanchez, W.; Bender, C.; Porcher, J.M. Wild gudgeons (Gobio gobio) from French rivers are contaminated by microplastics: Preliminary study and first evidence. *Environ. Res.* **2014**, *128*, 98–100. [CrossRef]
24. Güven, O.; Gökdağ, K.; Jovanović, B.; Kıdeyş, A.E. Microplastic litter composition of the Turkish territorial waters of the Mediterranean Sea, and its occurrence in the gastrointestinal tract of fish. *Environ. Pollut.* **2017**, *223*, 286–294. [CrossRef] [PubMed]
25. Biginagwa, F.; Mayoma, B.; Shashoua, Y.; Syberg, K.; Khan, F.R. First evidence of microplastics in the African Great Lakes: Recovery from Lake Victoria Nile perch and Nile tilapia. *J. Gt. Lakes Res.* **2016**, *42*, 1146–1149. [CrossRef]
26. Khan, F.R.; Mayoma, B.S.; Biginagwa, F.J.; Syberg, K. Microplastics in Inland African Waters: Presence, Sources, and Fate. In *Freshwater Microplastics. The Handbook of Environmental Chemistry*; Wagner, M., Lambert, S., Eds.; Springer: Heidelberg, Germany, 2018; Volume 58, pp. 101–124.
27. FishBase. Available online: https://www.fishbase.se/summary/Oreochromis-niloticus.html (accessed on 28 August 2019).

28. FishBase. Available online: https://www.fishbase.se/summary/5462 (accessed on 28 August 2019).
29. Egypt's Sherine Sentenced to Prison Over Nile Joke. Available online: https://www.bbc.com/news/world-middle-east-43217483 (accessed on 28 August 2019).
30. Basher, Z.; Lynch, A.J.; Taylor, W.W. New global high-resolution centerlines dataset of selected river systems. *Data Brief* **2018**, *20*, 1552–1555. [CrossRef] [PubMed]
31. Cole, M.; Webb, H.; Lindeque, P.K.; Fileman, E.S.; Halsband, C.; Galloway, T.S. Isolation of microplastics in biota-rich seawater samples and marine organisms. *Sci. Rep.* **2014**, *4*, 4528. [CrossRef] [PubMed]
32. Hidalgo-Ruz, V.; Gutow, L.; Thompson, R.C.; Thiel, M. Microplastics in the marine environment: A review of the methods used for identification and quantification. *Environ. Sci. Technol.* **2012**, *46*, 3060–3075. [CrossRef]
33. Nel, H.A.; Dalu, T.; Wasserman, R.J. Sinks and sources: Assessing microplastic abundance in river sediment and deposit feeders in an Austral temperate urban river system. *Sci. Total Environ.* **2018**, *612*, 950–956. [CrossRef]
34. Nor, N.H.; Obbard, J.P. Microplastics in Singapore's coastal mangrove ecosystems. *Mar. Pollut. Bull.* **2014**, *79*, 278–283.
35. Horton, A.A.; Jürgens, M.D.; Lahive, E.; van Bodegom, P.M.; Vijver, M.G. The influence of exposure and physiology on microplastic ingestion by the freshwater fish Rutilus rutilus (roach) in the River Thames, UK. *Environ. Pollut.* **2018**, *236*, 188–194. [CrossRef]
36. Tanaka, K.; Takada, H. Microplastic fragments and microbeads in digestive tracts of planktivorous fish from urban coastal waters. *Sci. Rep.* **2016**, *6*, 34351. [CrossRef]
37. Frias, J.P.; Otero, V.; Sobral, P. Evidence of microplastics in samples of zooplankton from Portuguese coastal waters. *Mar. Environ. Res.* **2014**, *95*, 89–95. [CrossRef]
38. Gago, J.; Galgani, F.; Maes, T.; Thompson, R.C. Microplastics in Seawater: Recommendations from the marine strategy framework directive implementation process. *Front. Mar. Sci* **2016**, *3*, 219. [CrossRef]
39. Mizraji, R.; Ahrendt, C.; Perez-Venegas, D.; Vargas, J.; Pulgar, J.; Aldana, M.; Ojeda, F.P.; Duarte, C.; Galbán-Malagón, C. Is the feeding type related with the content of microplastics in intertidal fish gut? *Mar. Pollut. Bull.* **2017**, *116*, 498–500. [CrossRef] [PubMed]
40. Foekema, E.M.; De Gruijter, C.; Mergia, M.T.; Murk, A.J.; van Franeker, J.A.; Koelmans, A.A. Plastic in North Sea fish. *Environ. Sci. Technol.* **2013**, *47*, 8818–8824. [CrossRef] [PubMed]
41. Rummel, C.D.; Löder, M.G.J.; Fricke, N.F.; Lang, T.; Griebeler, E.-M.; Janke, M.; Gerdts, G. Plastic ingestion by pelagic and demersal fish from the north sea and baltic sea. *Mar. Pollut. Bull.* **2016**, *102*, 134–141. [CrossRef]
42. Neves, D.; Sobral, P.; Ferreira, J.L.; Pereira, T. Ingestion of microplastics by commercial fish off the Portuguese coast. *Mar. Pollut. Bull.* **2015**, *101*, 119–126. [CrossRef]
43. Nadal, M.A.; Alomar, C.; Deudero, S. High levels of microplastic ingestion by the semipelagic fish bogue Boops boops (L.) around the Balearic Islands. *Environ. Pollut.* **2016**, *214*, 517–523. [CrossRef]
44. Bellas, J.; Martínez-Armental, J.; Martínez-Cámara, A.; Besada, V.; Martínez-Gomes, C. Ingestion of microplastics by demersal fish from the Spanish Atlantic and Mediterranean coasts. *Mar. Pollut. Bull.* **2016**, *109*, 55–60. [CrossRef]
45. Romeo, T.; Pietro, B.; Pedà, C.; Consoli, P.; Andaloro, F.; Fossi, M.C. First evidence of presence of plastic debris in stomach of large pelagic fish in the Mediterranean Sea. *Mar. Pollut. Bull.* **2015**, *95*, 358–361. [CrossRef]
46. Slootmaekers, B.; Carteny, C.C.; Belpaire, C.; Saverwyns, S.; Fremout, W.; Blust, R.; Bervoets, L. Microplastic contamination in gudgeons (Gobio gobio) from Flemish rivers (Belgium). *Environ. Pollut.* **2019**, *244*, 675–684. [CrossRef]
47. Pazos, R.S.; Maiztegui, T.; Colautti, D.C.; Paracampo, A.H.; Gómez, N. Microplastics in gut contents of coastal freshwater fish from Río de la Plata estuary. *Mar. Pollut. Bull.* **2017**, *122*, 85–90. [CrossRef]
48. Arias, A.H.; Ronda, A.C.; Oliva, A.L.; Marcovecchio, J.E. Evidence of microplastic ingestion by fish from the Bahía Blanca Estuary in Argentina, South America. *Bull. Environ. Contam. Toxicol.* **2019**, *102*, 750–756. [CrossRef] [PubMed]
49. Boerger, C.M.; Lattin, G.L.; Moore, S.L.; Moore, C.J. Plastic ingestion by planktivorous fish in the North Pacific Central Gyre. *Mar. Pollut. Bull.* **2010**, *60*, 2275–2278. [CrossRef] [PubMed]
50. Browne, M.A.; Galloway, T.S.; Thompson, R.C. Spatial patterns of plastic debris along estuarine shorelines. *Environ. Sci. Technol.* **2010**, *44*, 3404–3409. [CrossRef]
51. Bonhomme, S.; Cuer, A.; Delort, A.M.; Lemaire, J.; Sancelme, M.; Scott, G. Environmental biodegradation of polyethylene. *Polym. Degrad. Stab.* **2003**, *81*, 441–452. [CrossRef]

52. Velez, J.F.; Shashoua, Y.; Syberg, K.; Khan, F.R. Considerations on the use of equilibrium models for the characterisation of HOC-microplastic interactions in vector studies. *Chemosphere* **2018**, *210*, 359–365. [CrossRef]
53. Holmes, L.A.; Turner, A.; Thompson, R.C. Adsorption of trace metals to plastic resin pellets in the marine environment. *Environ. Pollut.* **2012**, *160*, 42–48. [CrossRef]
54. Song, Y.K.; Hong, S.H.; Jang, M.; Kang, J.H.; Kwon, O.Y.; Han, G.M.; Shim, W.J. Large accumulation of micro-sized synthetic polymer particles in the sea surface microlayer. *Environ. Sci. Technol.* **2014**, *48*, 9014–9021. [CrossRef]
55. Revel, M.; Châtel, A.; Mouneyrac, C. Micro (nano) plastics: A threat to human health? *Curr. Opin. Environ. Sci. Health* **2018**, *1*, 17–23. [CrossRef]
56. Akindele, E.O.; Ehlers, S.M.; Koop, J.H. First empirical study of freshwater microplastics in West Africa using gastropods from Nigeria as bioindicators. *Limnologica* **2019**, *78*, 125708. [CrossRef]
57. Toumi, H.; Abidli, S.; Bejaoui, M. Microplastics in freshwater environment: The first evaluation in sediments from seven water streams surrounding the lagoon of Bizerte (Northern Tunisia). *Environ. Sci. Pollut. Res.* **2019**, *26*, 14673–14682. [CrossRef]

Article

Effects of MP Polyethylene Microparticles on Microbiome and Inflammatory Response of Larval Zebrafish

Nicholas Kurchaba, Bryan J. Cassone, Caleb Northam, Bernadette F. Ardelli and Christophe M. R. LeMoine *

Department of Biology, Brandon University, Brandon, MB R7A 6A9, Canada; kurchan90@brandonu.ca (N.K.); cassoneb@brandonu.ca (B.J.C.); northamc@myumanitoba.ca (C.N.); ardellib@brandonu.ca (B.F.A.)

* Correspondence: lemoinec@brandonu.ca

Received: 8 July 2020; Accepted: 3 August 2020; Published: 11 August 2020

Abstract: Plastic polymers have quickly become one of the most abundant materials on Earth due to their low production cost and high versatility. Unfortunately, some of the discarded plastic can make its way into the environment and become fragmented into smaller microscopic particles, termed secondary microplastics (MP). In addition, primary MP, purposely manufactured microscopic plastic particles, can also make their way into our environment via various routes. Owing to their size and resilience, these MP can then be easily ingested by living organisms. The effect of MP particles on living organisms is suspected to have negative implications, especially during early development. In this study, we examined the effects of polyethylene MP ingestion for four and ten days of exposure starting at 5 days post-fertilization (dpf). In particular, we examined the effects of polyethylene MP exposure on resting metabolic rate, on gene expression of several inflammatory and oxidative stress linked genes, and on microbiome composition between treatments. Overall, we found no evidence of broad metabolic disturbances or inflammatory markers in MP-exposed fish for either period of time. However, there was a significant increase in the oxidative stress mediator L-FABP that occurred at 15 dpf. Furthermore, the microbiome was disrupted by MP exposure, with evidence of an increased abundance of *Bacteroidetes* in MP fish, a combination frequently found in intestinal pathologies. Thus, it appears that acute polyethylene MP exposure can increase oxidative stress and dysbiosis, which may render the animal more susceptible to diseases.

Keywords: microplastic; dysbiosis; microbiome; freshwater

---

1. Introduction

Plastic materials are inexpensively produced and provide a high level of durability and flexibility, making them a key material for various applications. While an increasing number of plastics are re-used and recycled, many "single use" plastics are readily discarded and accumulate as waste worldwide [1]. Over 4900 metric tons of plastic waste have been discarded in landfills and the environment since 1950 [1,2]. Discarded plastics are derived from different polymers, such as polyethylene, polypropylene, and polystyrene, with each polymer providing unique and desirable qualities. Polyethylene is particularly desirable for packaging due to its resistance to degradation and low economic cost and thus is of major environmental importance [3].

Plastic pollution affects most ecosystems, including freshwater and oceanic systems. In these environments, plastic litter undergoes photo-oxidation and physical breakdown, generating smaller and chemically stable particles. When these particles are smaller than 5 mm in diameter, they are called secondary microplastics (MP) and may persist extensively in the environment and in organisms [4–7]. In contrast, primary MP are intentionally manufactured for various applications [6]. Regardless of their provenance, primary and secondary MP have been found in oceans for decades, though more recently they have been detected in freshwater, reaching some of the most remote lakes around the world [2,8,9]. In freshwater reserves, such as the Ottawa River and the Nantaizi Lake, MP can attain concentrations between 0.1 and 6162.5 p/L [10–12]. As MP can easily be ingested, their presence in aquatic environments poses a potential risk to many animal species, chronically exposing them to this pervasive pollutant. In aquatic animals, MP primarily accumulate both in the alimentary canal and respiratory structures in adult animals, though they appear to be limited to the intestine in aquatic larvae [13,14].

Once ingested, MP can negatively impact organismal health at different levels [15–17]. First, MP can disrupt intestinal transit and their accumulation may trigger intestinal villi swelling [18]. In addition, at the cellular level, MP can also cause physical injury to surrounding cells via abrasion-induced oxidative stress and accumulation of intercellular reactive oxygen species (ROS) [19]. Enhanced intestinal ROS production may also indicate an inflammatory response and participate in a feedback loop [20]. Regardless of the cause, excess ROS production jeopardizes tissue integrity and overall organismal health [21].

The inflammatory response is coordinated by multiple cascades, including pro- and anti-inflammatory mediators. These cascades can be initiated by multiple stressors, such as toxins, tissue damage, or microorganisms [22], and may trigger cytokines to flood surrounding tissues and recruit nearby immune cells to initiate inflammation. While inflammation allows for the removal of the foreign stimuli and promotion of active healing of the injured area, prolonged inflammation can also result in various pathologies, such as inflammatory bowel disease, leading to metabolic dysfunction and nutrient deficiency [23,24].

The intestinal accumulation of MP may also affect its microbiome, another crucial player in animal wellbeing. In many organisms, including fish, it is clear that the microbiome plays a crucial role in both health and diseases; thus, disturbances in the gut microbiome can be detrimental to overall animal health [25–27]. As in other organisms, the fish microbiome can be affected by MP, with presumed widespread effects on animal health [28–33]. Indeed, gut dysbiosis has been repeatedly linked to various metabolic disorders [34–37].

Finally, the combined effect of oxidative stress, inflammation, and gut dysbiosis could have an even more pronounced impact on fish during early development when energy usage and physiological stress is high [38–40]. In particular, increased energy expenditure could negatively impact vital processes, such as tissue maintenance, locomotion, and growth of the individual. MP can also indirectly affect energy budgets by promoting an immune response to purge and remove plastic particles via inflammation [41].

In order to investigate the potential negative effects of MP on a freshwater fish model organism, we exposed larval zebrafish to polyethylene MP (10–40 μm) at a concentration of 20 mg/L. The range selected simulates a mixed-size exposure while providing an adequate representation of the type of microplastics found in the environment, and provides consistency with previous laboratory investigations using supra-ecological concentrations [42–45]. The experiment examined the acute and prolonged effects of MP exposure throughout early development by assessing oxygen consumption rates (OCRs), directed gene expression analysis, and microbiome diversity to assess the various impacts of differential MP exposure. Our results show that MP exposure during early life has no impact on metabolic or immunological systems but appears to have localized effects within the gut by increasing oxidative stress and disturbing the gut microbiome.

2. Methods

2.1. Caretaking

Commercially obtained adult zebrafish (*Danio rerio*) were maintained in an aquatic housing system (Aquaneering, San Diego, CA, USA) with recirculating UV-treated de-chlorinated water maintained at 27 °C with a 14:10 h light dark cycle. Adult animals were fed twice daily with Adult Zebrafish Complete Diet (Zeigler, Gardners, PA, USA) and once with *Artemia naupii*. Adult fish were bred in tanks using standard procedures [46], and embryos were collected before being washed twice in 0.05% bleach solution and transferred to 90 × 15 mm petri plates filled with E3 buffer solution (5 mM NaCl, 0.17 mM KCl, 0.33 mM CaCl2, 0.33 mM MgSO4, 0.0006 mM methylene blue). The fish were then kept at 28 °C until day 5 post-fertilization (5 dpf), with E3 medium being replaced once every 24 h. At 5 dpf, the fish were transferred into 300 mL beakers and maintained in de-chlorinated control water or water containing MP. The MP fish were exposed to commercially obtained 10–45 μm polyethylene microspheres (Cospheric, Goleta, CA, USA) at a concentration of 20 mg/L. All fish (MP and control) were kept under these conditions for either 4 or 10 days, corresponding to 9 dpf and 15 dpf, respectively. Under similar experimental conditions, this treatment previously yielded consistent ingestion and accumulation of MP in all zebrafish larvae [13]. All treatments were maintained in a 27 °C water bath for the duration of the experiment. Daily, larvae were fed ad libitum Larval AP100 feed (Ziegler, Gardners, PA, USA), and approximately one third of the water, 100 mL of either MP or pristine freshwater, was replaced to maintain water quality and remove deceased larvae. As with prior work [13], mortality rates remained low (<2%) throughout the experiment. Bioassays were run in parallel using fish from 6 independent breeding events. All animal experimentations conform to the guidelines of the Canadian Council on Animal Care and were approved by the Brandon University Animal Care Committee on July 12 2016 under the research protocol 2016R03.

2.2. Respirometry

At 9 and 15 dpf, the oxygen consumption rates of zebrafish larvae were measured using a fibre-optic respirometry system (OXY-4 mini, PreSens, Regensburg, Germany). Both oxygen consumption and temperature values were recorded in parallel with the AutoResp software v2.1.2 (Loligo® Systems, Viborg, Denmark), and data were analyzed using RespR v1.1.0.0 [47]. The respirometry procedure was conducted in a temperature-controlled water bath (26.91 °C ± 0.02) for the duration of the experiment. Fish were placed into borosilicate chambers (2–3 mL) fitted with small optodes using aquarium-grade silicon (Marineland, Blacksburg, VA, USA). Magnetic stir bars allowed for adequate oxygen circulation and their consistent rotation was improved by the addition of rubber rings at the bottom of the tube. The system was calibrated with 100% O_2 saturated E3 medium or dissolved sodium sulfite (0.02 g/mL solution). The fish were acclimated to the borosilicate chambers for 30 min and half of the total volume was replaced with fresh E3 prior to data collection. Each trial consisted of a 20 min sampling period featuring an early section (2 min), an intermediate section (8 min), and a late section (10 min). A sample size of 6 replicates (where each replicate included 5–10 fish depending on the size of the tube used) was used to determine OCRs between the treatments. We also measured background OCR (with no fish in chambers) and found these values to be negligible compared to larval OCRs. The intermediate sections were plotted using RespR to identify the most linear portion of the slope used for OCR calculation [47]. The OCRs were converted to nanomol/fish/hour to remain consistent with previous literature [13,48].

2.3. RNA Extraction, Reverse Transcription (RT), and qPCR

Following 4 or 10 days of exposure (9 dpf or 15 dpf), the larvae were anesthetized in 100 mg/L tricaine methanesulfonate (MS–222, Sigma-Aldrich, Darmstadt, Germany) and quick-frozen in liquid nitrogen before being stored at −80 °C until further manipulation. RNA was isolated from a total of 36 samples, comprising 9 replicates for control and MP treatments at both 9 and 15 dpf, using an RNeasy Mini extraction kit in accordance to the manufacturer's protocol (RNeasy, Qiagen, Aarhus, Denmark). A final volume of 20 μL was obtained for each sample, and the quantity and quality of the RNA was determined by the use of a NP80 NanoPhotometer® (Implen, Munich, Germany). Optical density ratios of 260/280 nm above 1.9 and concentrations above 500 ng/μL were targeted in order to provide a total of 2 μg of RNA for reverse transcription using the GoScript™ Reverse Transcription System

(Promega, Madison, WI, USA). The transcribed cDNA was then diluted (1:10) prior to qPCR, as described before [49].

2.4. qPCR

Nine genes were selected in order to examine the inflammatory response of larval zebrafish. Primer sets were obtained from previous publications or custom designed with the use of Primer3 software (Table 1, [50–57]). Primer specificity was confirmed by examining the top e-value in BLASTn (NCBI) and single product amplification confirmed by melt analysis of all reactions.

Table 1. List of primers used in qPCR assays of gene expression.

Gene	Forward Primer	Reverse Primer	Accession Number
ef1α[51]	CTTCTCAGGCTGACTGTGC	CCGCTAGCATTACCCTCC	AY422992
Rpl-13a[52]	TCTGGAGGACTGTAAGAGGTATGC	AGACGCACAATCTTGAGAGCAG	NM_212784.1
MyD88[53]	GAGGCGATTCCAGTAACAGC	GAAAGCATCAAAGGTCTCAGGTG	NM_212814.2
Ccl20[54]	ATCAATCTGCGCTAATCCATCAC	TGGTGAACATGCTCATCGTCTT	NM_001113595.1
L-FABP[55]	ACGTGGCAGGTTTACGCTCAG	TTGGAGGTGATGGTGAAGTCG	BC164928.1
SOD1[56]	GTCGTCTGGCTTGTGGAGTG	TGTCAGCGGGCTAGTGCTT	AY324390.1
NF-κβ[56]	CCAAATCCCAAAAGGTTAGAGATTT	CCTCTTAGGGCTGAGCGAATT	XM_005156814.2
Cxcl8a[57]	TGTTTTCCTGGCATTTCTGAC	TTTACAGTGTGGGCTTGGAGGG	XM_009306855.3
Mus2.1[50]	TGGTGGACCAGTGTGAAAAA	GGTCCAAAACCCAGCTACAA	XM_021470771.1

Gene expression analysis was carried by qPCR amplification using the QuantiNova™ SYBR Green PCR kit (Qiagen, Aarhus, Denmark), with a final volume of 15 μL per reaction (7.5 μL 2X SYBR Green PCR Master Mix, 0.7 μM forward and reverse primer, respectively, and 1 μL cDNA). Standard curves were created for each of the primer sets and the efficiency of each primer set was computed using Quantinova Rotor-Gene Q Series Software (Qiagen). Melting curve analysis was also performed to ensure single product amplification in each reaction. Only primer sets with suitable efficiency values (1.00 +/− 0.20) and single amplicon were used for further testing. The cycle threshold (Ct) values were automatically calculated using Rotor-Gene software using the corresponding standard curve. Relative gene expression levels were calculated using the ΔΔCt method [58], using the geometric mean of two housekeeping genes (*eF1α* and *Rpl-13a*) as a reference [49,58].

2.5. Respirometry and Gene Expression Statistical Analysis

Respirometry data and gene expression data were analysed in SigmaPlot software (SigmaPlot 14.0, Systat Software, San Jose, CA, USA). A two-way ANOVA was used to compare both the respirometry and gene expression data. All tests were performed at a significance level of $p = 0.05$, after passing equal variance and normality tests.

2.6. Microbiome Analysis

2.6.1. Sampling, DNA Extraction, Sequencing, and Pre-Processing

At both 9 and 15 dpf, 6 replicate samples of each treatment containing 20 individual zebrafish larvae each were anesthetized in 100 mg/L tricaine methanesulfonate, thoroughly washed with 100% ethanol, flash frozen, and stored at −80 °C until further analyses. For each sample, genomic DNA was isolated using the One-4-All Genomic DNA Miniprep Kit (BioBasic, Markham, ON, Canada).

Samples were then sent to Génome Québec Innovation Centre (McGill University, Montreal, QC, Canada) for quality control, PCR validation and reaction, amplicon barcoding, and normalization. The 16S primers used were 515F: GTGCCAGCMGCCGCGGTAA and 806R: GGACTACHVGGGTWTCTAAT [59]. A BioAnalyzer 2100 (Agilent Technologies, Santa Clara, CA, USA) was used to check the quality of amplicon libraries that were subsequently sequenced on the MiSeq platform in a paired-end 300 bp fashion. Raw sequence reads were deposited in the NCBI short sequence read archive (SRA) under the accession number PRJNA643286. The amplicons were adapter and quality trimmed in CLC Genomics Workbench v11.0.1 (Qiagen, Aarhus, Denmark; Quality Limit

= 0.00316, Minimum Length = 15, default parameters herein). The reads were then processed in CLC v11.0.1 for downstream analyses, as described previously [60].

2.6.2. CLC Analysis

We used the CLC Microbial Genomics Module 3.5 (Qiagen) pipeline for 16S analysis. First, operational taxonomic units (OTUs) were clustered at 97% similarity against the reference SILVA 132 database [61]. Chimeric sequences as well as OTUs detected in less than 0.05% of the total reads were then filtered out.

Bacterial communities were compared across ages and between control and MP fish. The OTU table was rarefied before OTU alpha diversity analysis was carried out using Shannon and Simpson's alpha diversity indices as well as total OTU in CLC. Differences between all groups were assessed with a non-parametric Kruskall–Wallis test followed by Mann–Whitney U pairwise tests ($\alpha = 0.05$).

Furthermore, beta diversity was evaluated with weighted and unweighted UNIFRAC as well as the Bray–Curtis index. Principal coordinates analyses (PCoA) of the distance plots were visualized, and multivariate analyses (PERMANOVA) were used to compare treatment differences. Differential abundance was computed using a binomial general linear model framework, and a likelihood test was used to detect differences across groups, followed by a Wald test for pairwise comparisons (FDR < 0.05).

3. Results

3.1. *Respirometry*

We explored the effects of MP exposure on routine metabolic rate using closed-system respirometry (Figure 1). Mean oxygen consumption rates (OCRs) between fish raised in clear water (control) and MP-exposed fish at both developmental time points were not statistically significant (Figure 1). However, we were able to identify a difference in OCRs when comparing the two time points, 9 dpf and 15 dpf ($p = 0.015$). To test whether the developmental time point of the zebrafish was more important than the overall MP duration, we also examined a 10–15 dpf exposure period to MP; however, no significant difference was found when compared to the 5–15 dpf treatment (data not shown). Ultimately, no statistically significant difference in mean OCRs was identified in any of the treatments in regard to MP exposure (Figure 1).

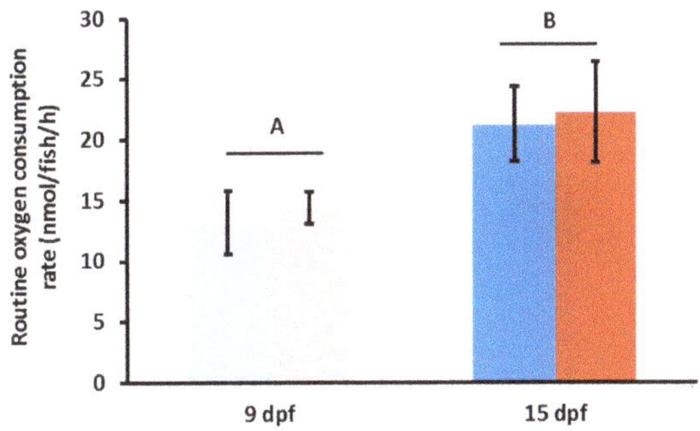

Figure 1. Oxygen consumption rates (OCRs) for microplastic (MP) exposed zebrafish. OCRs of MP fish after 5 days (pink) and 10 days (red) of exposure to MP and the respective age matched control fish (light blue/dark blue). No statistical significance was observed between the MP and control groups; however, there was a significant difference between time points ($n = 6$, $p = 0.015$).

3.2. *Gene Expression Analysis*

In order to assess the effects of MP exposure on various cellular response pathways, we used qPCR and evaluated patterns of gene expression of nine gene targets involved with the inflammatory response, the gut microbiome, or oxidative stress (Figure 2). In terms of MP inflammatory response, none of the target genes (*ccl20*, *cxcl8a*, and *NF-κβ*) appeared to be affected across treatments. Although *ccl20* and *cxcl8a* both show a marginal reduction in gene expression in 15 dpf MP fish, this effect was not statistically significant (Figure 2A). Similarly, genes associated with the gut microbiome (*MyD88* and *Mus2.1*) exhibited relatively high variability resulting in no appreciable difference between treatments (Figure 2B). Finally, genes that are typically upregulated in periods of oxidative stress, *L-FABP* and *SOD1*, were also examined. In this case, we detected a significant increase in *L-FABP* expression in MP fish at day 15 ($p = 0.022$), but a marginal and insignificant increase in *SOD1* in the same fish (Figure 2C).

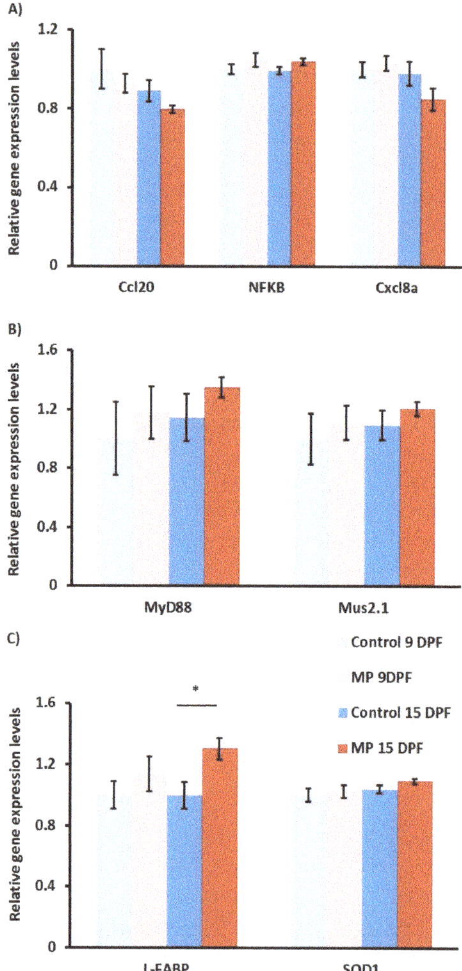

Figure 2. Gene expression in MP and control fish at 9 and 15 dpf. Gene expression analysis of inflammatory (**A**), microbial (**B**), and oxidative stress (**C**) genes in fish exposed to MP for 4 (pink) and 10 days (red) or aged matched control fish (light and dark blue). *EF1α/RLP-13a* were used as housekeeping genes. Statistically significant differences are indicated with an asterisk ($n = 9$, $p < 0.05$).

3.3. Structure of Gut Bacteria Communities

We aimed to characterize the effect of MP exposure on the microbiome of zebrafish larvae for different exposure duration. Fish were exposed to MP or control conditions and sampled at 4 and 10 days, and their

DNA was extracted ($n = 6$ for each treatment and time point) and was subjected to high throughput 16S sequencing. We obtained a total of 3,535,330 reads, of which 42% were retained as OTU reads after quality trimming and chimeric read removal. CLC analysis assigned these reads to 1239 OTUs at >97% similarity, and after selecting OTUs present in over 0.05% ($n = 105$) of the reads, these were identified from 7 phyla, 9 classes, 24 orders, 33 families, and 62 genera (see Supplementary File 1).

3.4. Core Microbial Composition

In this study, we describe the core microbiome as taxa that are present in all 24 of the samples sequenced, regardless of treatment and time point. At the phylum level, *Proteobacteria* had a consistent dominating presence as the most abundant phylum (90%) across all samples and treatments. Indeed, the next most abundant ubiquitous phyla, *Bacteroidetes* and *Cyanobacteria*, only each averaged 4% of the total reads in all samples (Figure 3). At the genus level, *Aeromonas*, *Pseudomonas*, and *Vibrio* were overall the most abundant and were among the 17 genera identified as the core microbiome common (Supplementary File 1). Thus, while there were some marked differences in microbial composition across samples and treatments (see below), there was a relatively strong presence of two major phyla and several genera in the larval zebrafish microbiome.

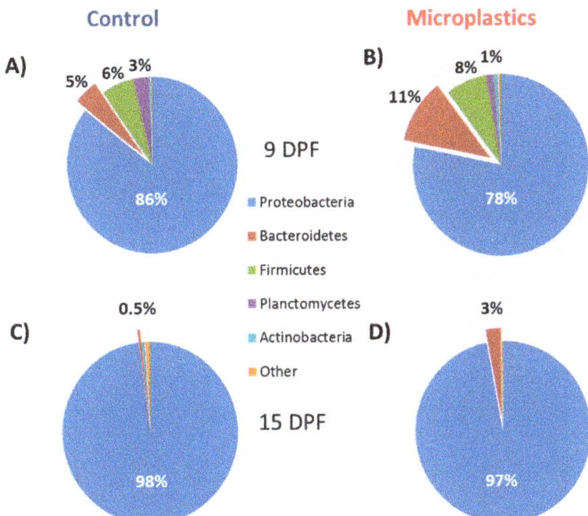

Figure 3. Relative phylum composition of control and MP-exposed zebrafish larvae. Zebrafish larvae were exposed to control conditions (**A**,**C**) and microplastics (**B**,**D**) for 4 days (**A**,**B**) and 10 days (**C**,**D**), respectively. Percentages represent the relative abundance of the most represented phyla.

3.5. Microbial Community Changes through Ontogeny

In order to assess the differences in microbiota across treatments, we assessed the relative diversity of the samples. First, we normalized for sequencing depth and rarified all samples to 37,895 reads to examine changes in community structure. Overall, as the fish aged, we saw a relative decrease in community diversity (Figure 4). Although the general pattern applies to both MP and control treatments, it was only statistically significant for all three indicators used in MP fish (observed OTUs, Shannon and Simpson indices), while only the total observed OTU index was significantly decreased in older control fish. However, when we delve into the specifics, we see the number of ubiquitous phyla decreasing from five to only two, *Bacteriodetes* and *Proteobacteria*, during the 9 to 15 dpf transition (Supplementary File 1). Incidentally, the abundance of these two phyla increased with the age of the fish, while others, such as *Firmicutes* and *Planctomycetes*, were reduced in older fish. These trends were further confirmed as several genera of the *Burkholderiaceae* and *Comamonadaceae* family, for example, markedly decreased in older fish (Supplementary File 1). Overall, there was strong evidence of an ontogenic shift in microbial fauna in the developing fish.

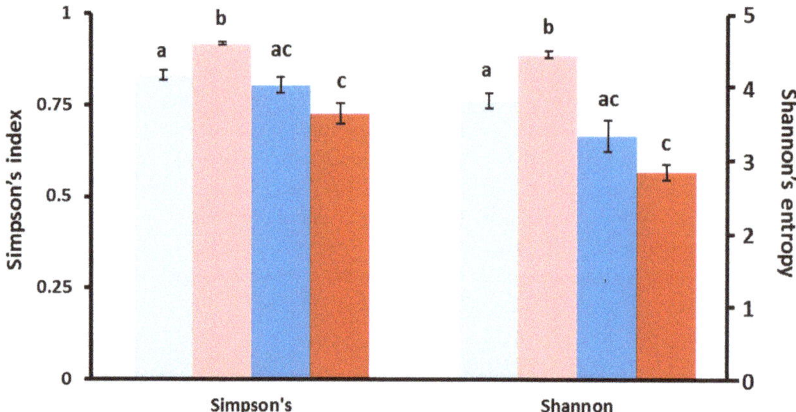

Figure 4. The effect of MP exposure on diversity indices in larval zebrafish. Simpson and Shannon OTU diversity indices obtained through CLC analysis for 9 dpf control (light blue) and microplastics (pink), as well as 15 dpf control (dark blue) and microplastics (red). Significant differences are indicated by different letters ($p = 0.05$).

3.6. MP Exposure Disturbs the Microbiome

While the age of the fish appears to have an influence on the microbiota composition within each treatment group, we were mostly interested in the differences within and between treatments. We compared beta diversity differences among our samples using permutational multivariate analysis (PERMANOVA). Overall, microbial communities were different across treatments/exposures (PERMANOVA: $F = 6.16$, $p = 3 \times 10^{-5}$) and particularly according to the age of the fish (PERMANOVA: $p = 3 \times 10^{-5}$). However, there was no overall significant impact of MP treatment compared to control when combining both time points (PERMANOVA: $F = 1.11$, $p = 0.342$). When examined within age groups, only 9 dpf fish showed a significant difference in bacterial communities in MP-treated fish compared to controls ($p = 0.013$). However, at both time points, there was an increased abundance of *Bacteroidetes* in MP fish, as well as changes in the abundance of several genera (Supplementary File 1). This overall suggests that, while larval zebrafish undergo broad ontogenic microbial changes over the exposure period, there is an evident effect of MP on the microbiome of larval fish, particularly during early development.

4. Discussion

Plastic pollution is quickly becoming one of the most concerning forms of anthropogenic pollution, with numerous studies indicating increased levels of MP in our water systems [62–64]. MP can accumulate inside aquatic species, rapidly leading to bioaccumulation and homeostatic imbalances [65,66]. For example, wild fish are likely encountering MP chronically and thus serve as excellent models for the study of MP toxicity in a living system [67]. A number of studies have demonstrated that MP pollution is indeed harmful, even under acute exposures, with polystyrene MP causing gut dysbiosis, inflammation, oxidative stress, and impaired swimming performance in adult zebrafish [41,68]. However, exposure to polyethylene MP has shown conflicting results in adult zebrafish, with some suggesting that polyethylene MP pose little to no harm [69], while others describe signs of oxidative stress and inflammation [16]. Here, we present additional evidence that there is a limited impact of acute MP exposure on developing larval zebrafish. Indeed, ten days of exposure to supra-ecological concentrations of MP does not seem to affect the overall metabolism of the fish and has a limited impact on inflammatory molecular responses, but it seems to trigger an elevated ROS response in the larvae and a concurrent dysbiosis of the larval microbiome.

In order to assess the effect of MP ingestion at the whole organism level, we monitored larval oxygen consumption in a closed respirometry system. Indeed, changes in oxygen consumption could provide an indication of changes in the metabolic status of the animal, including changes in immunological activity and physiological stress [70,71]. Overall, larval oxygen consumption rates were in line with previous studies, with an evident increase in oxygen consumption as the larvae ages [48]. However, we were unable to detect any differences in oxygen consumption between age-matched control and MP-exposed fish, suggesting no global effects on the larval metabolic function. Increased OCRs may indicate the presence of physiological stressors such as widespread inflammation or decreased nutrient absorption [72]. Although similar oxygen consumption rates could indicate a lack of response, we cannot rule out a localized immune response to MP in the fish which may not have broad metabolic effects [13,44]. To further investigate this possibility, we examined the expression of several immune response genes following distinct durations of MP exposure.

Early on in larval zebrafish development, the immune system relies entirely on the innate response for protection [73]. Though efficient, this system requires activation via several pattern recognition receptors that recognize specific exogenous molecules and trigger the release of inflammatory cytokines [22]. We hypothesized that these receptors could be activated by polyethylene particles and thus decided to examine the expression of two genes involved in the broad spectrum *MyD88*-dependent pathway. *MyD88* is largely responsible for the signal transmission of toll-like receptors associated with inflammation, ultimately upregulating cytokine *NF-Kβ* [74–76]. We were unable to find any significant difference in *MyD88* or *NF-Kβ* expression between either of the time points and treatments, suggesting an absence of generalized inflammation via this pathway. However, it is possible that, given the broad tissue distribution of these markers, our approach measuring whole body gene expression levels may mask small localized changes in intestinal tissues. Considering this caveat, we investigated two cytokines (*ccl20* and *cxcl8a*) closely associated with the gut region to provide a more targeted assessment. Both cytokines have been associated with localized gut inflammation in zebrafish in various contexts, which may identify them as key contributors to gut inflammation [77,78]. However, once again, we did not detect any differences in gene expression, further suggesting that intestinal inflammation does not occur in larval zebrafish exposed to MP.

Despite the apparent lack of inflammation, we suspected that MP in the gut could lead to increased oxidative stress resulting from physical abrasion of enterocytes. Therefore, we examined the transcript accumulation of two intestinal antioxidants found in the intestine of larval zebrafish: liver fatty-acid binding protein (*L-FABP*) and superoxide dismutase 1 (*SOD1*). *L-FABP* mitigates hepatic and intestinal ROS production by binding to fatty acid metabolites and heme ligands [79]. Similarly, *SOD1* limits ROS using zinc and copper ions, producing non-reactive hydrogen and oxygen atoms [80]. We observed an increase in *L-FABP* expression during the MP 9–15 dpf time period; however, there was no observable difference in *SOD1* expression, possibly due to the wide expression of *SOD1* in contrast to a more tissue-specific expression of *L-FABP*. Together, as previously suggested, our results suggest that MP exposure promotes increased ROS in the zebrafish gut [16,28,81].

Microbiome

The gut of the zebrafish is quickly colonized by microorganisms during the early phases of larval development following the completion of the intestinal tract, resulting in a microbial community known as the "microbiome". The microbiome is primarily colonized by bacteria that originate from the chorion of the larvae and the surrounding environment during the initial feeding event [82]. Consequently, variability in environmental conditions has made it complex to assess the composition of a core intestinal microbiome in zebrafish, particularly when considering the rapid development of these fish and the variability among fish facilities and strains [83,84]. With these aspects in mind, we ensured that zebrafish larvae were treated equally during our two periods of MP exposure (5–9 dpf and 5–15 dpf) to identify potential time-sensitive responses to polyethylene MP.

As the fish aged, the diversity of the microbiome was significantly reduced. Furthermore, trends of dominant genera changed from 9 dpf to 15 dpf in both the absence and presence of MP. In all fish, dominant genera such as *Variovorax* increased with age, while others such as *Rheinheimera* disappeared. When comparing the endpoints of both treatments, we were able to identify a decrease in *Aeromonas* sp. and an increase in *Vibrio* sp., a change which may be indicative of natural microbiome development [85]. Increased microbiome diversity is expected to occur at approximately 10 dpf as the zebrafish larvae transition from larvae to juveniles, meaning that this period of development is naturally highly dynamic, even in the absence of MP or other contaminants [84]. We were able to detect two distinct microbial community profiles between half of our samples in each of the MP treatments, which suggests that our time points represent this transitional state and may have increased the susceptibility to dysbiosis via MP exposure.

Larval microbiota was overall more affected by MP at 9 dpf than 15 dpf, indicating that dysbiosis in response to this pollutant may be time- or life stage-sensitive. In contrast to older phases of development, larval zebrafish seem to experience greater gut microbial shifts due to environmental factors [84,86,87]. However, at the phylum level, MP induced an increase in *Bacteroidetes* at both life-stages and a decrease in *Actinobacteria* at 15 dpf, a combination that has been found in intestinal inflammation pathologies [88]. Furthermore, some members of the sub-genera Bacteroids have also had their presence associated with various human metabolic diseases such as obesity, autism, type II diabetes, and colorectal cancer [34–36,89]. Therefore, although MP exposure only transiently affects bacterial communities, these changes could negatively impact the health of the fish during these transitional life stages.

Interestingly, 15 dpf MP larvae demonstrate concurrently a higher abundance of *Bacteriodetes* and an increase in ROS, as evidenced by an increase in *L-FABP* expression compared to control fish at 15 dpf. While we can only speculate regarding the interaction between ROS production and increased abundance of this bacterial phylum, it is possible that increased oxidative stress in MP fish promotes the growth of these bacteria. Indeed, while cells of many micro and macro-organisms are typically affected by ROS [68,90,91], some microbial species, including members of the *Bacteriodetes*, have effective defense mechanisms against damaging oxidative stress, which may give them a selective advantage over other micro-organisms [92–94]. Alternatively, it is possible that the proliferation of these bacteria may be the underlying cause of oxidative stress. Regardless of the proximate causal agent, we confirm that MP exposure is detrimental to larval zebrafish.

5. Conclusions

MP exposure is an environmental concern, as virtually every single ecosystem is plagued with these stable microscopic particles. Here, we confirm that MP exposure negatively affects a freshwater organism as it can result in microbial dysbiosis and increased oxidative stress in larval zebrafish. These effects were asynchronous, with pronounced dysbiosis after only four days of exposure and a subsequent increase in ROS after longer exposure. Furthermore, we were unable to identify any determinants of metabolic distress or inflammation across three independent time points, which suggests that, during early development, MP toxicity may be localized (guts) and does not occur at the whole organism level. Our time course also suggests that the zebrafish gut is most susceptible to MP-induced dysbiosis following the transition from the larval to juvenile state and that, by promoting changes in the microbial communities, such as the proliferation of potentially pathogenic bacteria, MP has the potential to negatively affect fish health at a particularly vulnerable phase of development. Overall, our results corroborate the accumulating evidence that aquatic organisms are negatively affected by MP exposure, though the consequences at the tissue and whole organismal level remain complex and should certainly remain an active area of research.

Supplementary Materials: The following are available online at http://www.mdpi.com/2305-6304/8/3/55/s1, file 1: Differential microbial abundance after 4 and 10 days of MP exposure in larval zebrafish.

Author Contributions: Conceptualization, C.M.R.L.; Formal analysis, N.K. and C.M.R.L.; Funding acquisition, B.J.C. and C.M.R.L.; Investigation, N.K. and C.N.; Methodology, N.K. and B.F.A.; Supervision, B.J.C., B.F.A. and C.M.R.L.; Writing—Original draft, N.K. and C.M.R.L.; Writing—Review & editing, N.K., B.J.C. and B.F.A. All authors have read and agreed to the published version of the manuscript.

Funding: This study was supported by individual NSERC funding to C.M.R.L., B.A. and B.J.C., as well as collaborative CFI funding to C.M.R.L. and B.J.C.

Acknowledgments: We thank O.E. Elebute for guidance in the microbiome analysis.

Conflicts of Interest: The authors declare no conflict of interest.

References

1. Geyer, R.; Jambeck, J.R.; Law, L.K. Production, Use, and Fate of All Plastics Ever Made. *Sci. Adv.* **2017**, *3*, e1700782. [CrossRef] [PubMed]
2. Barnes, D.K.A.; Galgani, F.; Thompson, C.R.; Barlaz, M. Accumulation and Fragmentation of Plastic Debris in Global Environments. *Philos. Trans. R. Soc. Lond. B. Biol. Sci.* **2009**, *364*, 1985–1998. [CrossRef] [PubMed]
3. Roy, P.K.; Hakkarainen, M.; Varma, I.K.; Albertsson, A.C. Degradable Polyethylene: Fantasy or Reality. *Environ. Sci. Technol.* **2011**, *45*, 4217–4227. [CrossRef] [PubMed]
4. Andrady, A.L. Microplastics in the Marine Environment. *Mar. Pollut. Bull.* **2011**, *62*, 1596–1605. [CrossRef] [PubMed]
5. Cheshire, A.; Ellik, A.; Barbière, J.; Cohen, Y.; Evans, S.; Jarayabhand, S.; Jeftic, L.; Jung, R.T.; Kinsey, S.; Kusui, E.T.; et al. *UNEP/IOC Guidelines on Survey and Monitoring of Marine Litter*; Regional Seas Reports and Studies No. 186 IOC Technical Series No. 83. UNEP Regional Seas Reports and Studies; Intergovernmental Oceanographic Commission: Paris, France, 2009.
6. Frias, J.P.; Nash, R. Microplastics: Finding a consensus on the definition. *Mar. Pollut. Bull.* **2019**, *138*, 145–147. [CrossRef] [PubMed]
7. Hidalgo, R.V.; Gutow, L.; Thompson, R.C.; Thiel, M. Microplastics in the Marine Environment: A Review of the Methods Used for Identification and Quantification. *Environ. Sci. Technol.* **2012**, *46*, 3060–3075. [CrossRef]
8. Obbard, R.W.; Sadri, S.; Wong, Q.Y.; Khitun, A.A.; Baker, I.; Thompson, R.C. Global Warming Releases Microplastic Legacy Frozen in Arctic Sea Ice. *Earths Future* **2014**, *2*, 315–320. [CrossRef]
9. Sun, J.; Dai, X.; Wang, Q.; Loosdrecht, V.M.C.M.; Ni, B.J. Microplastics in Wastewater Treatment Plants: Detection, Occurrence and Removal. *Water Res.* **2019**, *152*, 21–37. [CrossRef]
10. Li, J.; Liu, H.; Chen, J.P. Microplastics in Freshwater Systems: A Review on Occurrence, Environmental Effects, and Methods for Microplastics Detection. *Water Res.* **2018**, *137*, 362–374. [CrossRef]
11. Shahul, H.F.; Bhatti, M.S.; Anuar, N.; Anuar, N.; Mohan, P.; Periathamby, A. Worldwide Distribution and Abundance of Microplastic: How Dire Is the Situation? *Waste Manag. Res.* **2018**, *36*, 873–897. [CrossRef]
12. Vermaire, C.J.; Pomeroy, C.; Herczegh, M.S.; Haggart, O.; Murphy, M. Microplastic Abundance and Distribution in the Open Water and Sediment of the Ottawa River, Canada, and Its Tributaries. *FACETS.* **2017**, *2*, 301–314. [CrossRef]
13. LeMoine, C.M.R.; Kelleher, B.M.; Lagarde, R.; Northam, C.; Elebute, O.O.; Cassone, B.J. Transcriptional Effects of Polyethylene Microplastics Ingestion in Developing Zebrafish (*Danio rerio*). *Environ. Pollut.* **2018**, *243 Pt A*, 591–600. [CrossRef]
14. Wright, S.L.; Thompson, R.C.; Galloway, T.S. The Physical Impacts of Microplastics on Marine Organisms: A Review. *Environ. Pollut.* **2013**, *178*, 483–492. [CrossRef] [PubMed]
15. Chambers, T.K.; Chen, Z.C.; Crawford, P.A.; Fu, X.; Burgess, S.C.; Lai, L.; Leone, T.C.; Kelly, D.P.; Finck, B.N. Liver-Specific PGC-1beta Deficiency Leads to Impaired Mitochondrial Function and Lipogenic Response to Fasting-Refeeding. *PLoS ONE* **2012**, *7*, e52645. [CrossRef] [PubMed]

16. Lu, Y.; Zhang, Y.; Deng, Y.; Jiang, W.; Zhao, Y.; Geng, J.; Ding, L.; Ren, H. Uptake and Accumulation of Polystyrene Microplastics in Zebrafish (*Danio rerio*) and Toxic Effects in Liver. *Environ. Sci. Technol.* **2016**, *50*, 4054–4060. [CrossRef]
17. Pitt, J.A.; Kozal, J.S.; Jayasundara, N.; Massarsky, A.; Trevisan, R.; Geitner, N.; Wiesner, M.; Levin, E.D.; Di Giulio, R.T. Uptake, Tissue Distribution, and Toxicity of Polystyrene Nanoparticles in Developing Zebrafish (*Danio rerio*). *Aquat. Toxicol.* **2018**, *194*, 185–194. [CrossRef]
18. Pedà, C.; Leteria, C.; Fossi, C.M.; Gai, F.; Andaloro, F.; Genovese, L.; Perdichizzi, A.; Romeo, T.; Maricchiolo, G. Intestinal Alterations in European Sea Bass *Dicentrarchus labrax* (Linnaeus, 1758) Exposed to Microplastics: Preliminary Results. *Environ. Pollut.* **2016**, *212*, 251–256. [CrossRef]
19. Jeong, C.B.; Won, E.J.; Kang, H.-M.; Lee, M.C.; Hwang, D.S.; Hwang, U.K.; Zhou, B.; Souissi, S.; Lee, S.J.; Lee, J.S. Microplastic Size-Dependent Toxicity, Oxidative Stress Induction, and p-JNK and p-P38 Activation in the Monogonont Rotifer (*Brachionus koreanus*). *Environ. Sci. Technol.* **2016**, *50*, 8849–8857. [CrossRef]
20. Blaser, H.; Dostert, C.; Mak, T.W.; Brenner, D. TNF and ROS crosstalk in inflammation. *Trends Cell Biol.* **2016**, *26*, 249–261. [CrossRef]
21. Duan, X.D.; Feng, L.; Jiang, W.D.; Wu, P.; Liu, Y.; Kuang, S.Y.; Tang, T.; Tang, W.N.; Zhang, Y.A.; Zhou, X.Q. Dietary soybean β-conglycinin suppresses growth performance and inconsistently triggers apoptosis in the intestine of juvenile grass carp (*Ctenopharyngodon idella*) in association with ROS-mediated MAPK signalling. *Aquac. Nutr.* **2019**, *25*, 770–782. [CrossRef]
22. Chen, L.; Deng, H.; Cui, H.; Fang, C.; Zuo, Z.; Deng, J.; Li, Y.; Wang, X.; Zhao, L. Inflammatory Responses and Inflammation-Associated Diseases in Organs. *Oncotarget* **2018**, *9*, 7204–7218. [CrossRef] [PubMed]
23. Kaser, A.; Tilg, H. "Metabolic Aspects" in Inflammatory Bowel Diseases. *Curr. Drug Deliv.* **2012**, *9*, 326–332. [CrossRef] [PubMed]
24. Weisshof, R.; Chermesh, I. Micronutrient Deficiencies in Inflammatory Bowel Disease. *Curr. Opin. Clin. Nutr. Metab. Care* **2015**, *18*, 576–581. [CrossRef] [PubMed]
25. Slyepchenko, A.; Maes, M.; Machado-Vieira, R.; Anderson, G.; Solmi, M.; Sanz, Y.; Berk, M.; Köhler, C.A.; Carvalho, A.F. Intestinal dysbiosis, gut hyperpermeability and bacterial translocation: Missing links between depression, obesity and type 2 diabetes. *Curr. Pharm Des.* **2016**, *22*, 6087–6106. [CrossRef] [PubMed]
26. Lynch, S.V.; Pedersen, O. The human intestinal microbiome in health and disease. *N. Engl. J. Med.* **2016**, *375*, 2369–2379. [CrossRef]
27. Kim, A.; Kim, N.; Roh, J.H.; Chun, W.-K.; Ho, D.T.; Lee, Y.; Kim, D.-H. Administration of antibiotics can cause dysbiosis in fish gut. *Aquaculture* **2019**, *512*, 734330. [CrossRef]
28. Qiao, R.; Cheng, S.; Lu, Y.; Zhang, Y.; Ren, H.; Lemos, B. Microplastics Induce Intestinal Inflammation, Oxidative Stress, and Disorders of Metabolome and Microbiome in Zebrafish. *Sci. Total Environ.* **2019**, *662*, 246–253. [CrossRef]
29. Wan, Z.; Wang, C.; Zhou, J.; Shen, M.; Wang, X.; Fu, Z.; Jin, Y. Effects of polystyrene microplastics on the composition of the microbiome and metabolism in larval zebrafish. *Chemosphere* **2019**, *217*, 646–658. [CrossRef]
30. Fackelmann, G.; Sommer, S. Microplastics and the gut microbiome: How chronically exposed species may suffer from gut dysbiosis. *Mar. Pollut. Bull.* **2019**, *143*, 193–203. [CrossRef]
31. Lu, L.; Luo, T.; Zhao, Y.; Cai, C.; Fu, Z.; Jin, Y. Interaction between microplastics and microorganism as well as gut microbiota: A consideration on environmental animal and human health. *Sci. Total Environ.* **2019**, *667*, 94–100. [CrossRef]
32. Luo, T.; Wang, C.; Pan, Z.; Jin, C.; Fu, Z.; Jin, Y. Maternal polystyrene microplastic exposure during gestation and lactation altered metabolic homeostasis in the dams and their F1 and F2 offspring. *Environ. Sci. Technol.* **2019**, *53*, 10978–10992. [CrossRef] [PubMed]
33. Zhu, D.; Chen, Q.-L.; An, X.-L.; Yang, X.-R.; Christie, P.; Ke, X.; Wu, L.H.; Zhu, Y.G. Exposure of soil collembolans to microplastics perturbs their gut microbiota and alters their isotopic composition. *Soil Biol. Biochem.* **2018**, *116*, 302–310. [CrossRef]
34. Sobhani, I.; Tap, J.; Roudot-Thoraval, F.; Roperch, J.P.; Letulle, S.; Langella, P.; Corthier, G.; Nhieu, J.T.V.; Furet, J.P. Microbial Dysbiosis in Colorectal Cancer (CRC) Patients. *PLoS ONE* **2011**, *6*, e16393. [CrossRef] [PubMed]

35. Walker, A.W.; Sanderson, J.D.; Churcher, C.; Parkes, G.C.; Hudspith, B.N.; Rayment, N.; Brostoff, J.; Parkhill, J.; Gordon, D.; Petrovska, L. High-Throughput Clone Library Analysis of the Mucosa-Associated Microbiota Reveals Dysbiosis and Differences between Inflamed and Non-Inflamed Regions of the Intestine in Inflammatory Bowel Disease. *BMC Microbiol.* **2011**, *11*, 7. [CrossRef] [PubMed]
36. Wu, X.; Ma, C.; Han, L.; Nawaz, M.; Gao, F.; Zhang, X.; Yu, P.; Zhao, C.; Li, L.; Zhou, A.; et al. Molecular Characterisation of the Faecal Microbiota in Patients with Type II Diabetes. *Curr. Microbiol.* **2010**, *61*, 69–78. [CrossRef]
37. Reese, A.T.; Cho, E.H.; Kiltzman, B.; Nichols, S.P.; Wisniewski, N.A.; Villa, M.M.; Durand, H.K.; Jiang, S.; Midani, F.S.; Nimmagadda, S.N.; et al. Antibiotic-Induced Changes in the Microbiota Disrupt Redox Dynamics in the Gut. *eLife* **2018**, *7*, e35987. [CrossRef]
38. Gil, D.; Alfonso, I.S.; Pérez, R.L.; Muriel, J.; Monclús, R. Harsh Conditions during Early Development Influence Telomere Length in an Altricial Passerine: Links with Oxidative Stress and Corticosteroids. *J. Evol. Biol.* **2019**, *32*, 111–125. [CrossRef]
39. Jiang, N.M.; Cowan, M.; Moonah, S.N.; Petri, W.A. The Impact of Systemic Inflammation on Neurodevelopment. *Trends Mol. Med.* **2018**, *24*, 794–804. [CrossRef]
40. Mead, K.S.; Denny, M.W. The Effects of Hydrodynamic Shear Stress on Fertilization and Early Development of the Purple Sea Urchin *Strongylocentrotus purpuratus*. *Biol. Bull.* **1995**, *188*, 46–56. [CrossRef]
41. Jin, Y.; Xia, J.; Pan, Z.; Yang, J.; Wang, W.; Fu, Z. Polystyrene Microplastics Induce Microbiota Dysbiosis and Inflammation in the Gut of Adult Zebrafish. *Environ. Pollut.* **2018**, *235*, 322–329. [CrossRef]
42. Song, Y.K.; Hong, H.S.; Jang, M.; Kang, J.-H.; Kwon, O.Y.; Han, G.M.; Shim, W.J. Large Accumulation of Micro-Sized Synthetic Polymer Particles in the Sea Surface Microlayer. *Environ. Sci. Technol.* **2014**, *48*, 9014–9021. [CrossRef] [PubMed]
43. Watts, A.J.R.; Lewis, C.; Goodhead, R.M.; Beckett, S.J.; Moger, J.; Tyler, C.R.; Galloway, T.S. Uptake and retention of microplastics by the shore crab *Carcinus maenas*. *Environ. Sci. Technol.* **2014**, *48*, 8823e8830. [CrossRef] [PubMed]
44. Cole, M.; Lindeque, P.; Fileman, E.; Halsband, C.; Galloway, T.S. The Impact of Polystyrene Microplastics on Feeding, Function and Fecundity in the Marine Copepod Calanus Helgolandicus. *Environ. Sci. Technol.* **2015**, *49*, 1130–1137. [CrossRef] [PubMed]
45. Khan, R.F.; Syberg, K.; Shashoua, Y. Influence of polyethylene microplastic beads on the uptake and localization of silver in zebrafish (Danio rerio). *Environ. Pollut.* **2015**, *206*, 73e79. [CrossRef] [PubMed]
46. Westerfield, M. *The Zebrafish Book*; University of Oregon Press: Eugene, OR, USA, 1995.
47. Harianto, J.; Carey, N.; Byrne, M. respR—An R package for the manipulation and analysis of respirometry data. *Methods Ecol. Evol.* **2019**, *10*, 912–920. [CrossRef]
48. Rombough, P.; Drader, H. Hemoglobin Enhances Oxygen Uptake in Larval Zebrafish (*Danio rerio*) but Only under Conditions of Extreme Hypoxia. *J. Exp. Biol.* **2009**, *212 Pt 6*, 778–784. [CrossRef]
49. Northam, C.; LeMoine, C.M.R. Metabolic Regulation by the PGC-1α and PGC-1β Coactivators in Larval Zebrafish (*Danio rerio*). *Comp. Biochem. Physiol. A Mol. Integr. Physiol.* **2019**, *234*, 60–67. [CrossRef]
50. Untergasser, A.S.; Cutcutache, I.; Koressaar, T.; Ye, J.; Faircloth, B.C.; Remm, M.; Rozen, S.G. Primer3–new capabilities and interfaces. *Nucleic Acids Res.* **2012**, *40*, e115. [CrossRef]
51. McCurley, A.T.; Callard, G.V. Characterization of housekeeping genes in zebrafish: Male-female differences and effects of tissue type, developmental stage and chemical treatment. *BMC Mol. Biol.* **2008**, *9*, 102. [CrossRef]
52. Ibarra, M.S.; Etichetti, C.B.; Benedetto, C.D.; Rosano, G.L.; Margarit, E.; Sal, G.D.; Mione, M.; Girardini, J. Dynamic Regulation of Pin1 Expression and Function during Zebrafish Development. *PLoS ONE* **2017**, *12*, e0175939. [CrossRef]
53. Vaart, V.D.M.; Soest, V.J.J.; Spaink, H.P.; Meijer, A.H. Functional analysis of a zebrafish myd88 mutant identifies key transcriptional components of the innate immune system. *Dis. Model Mech.* **2013**, *6*, 841–854. [CrossRef]
54. Rotman, J.; Gils, V.W.; Butler, D.; Spaink, H.P.; Meijer, A.H. Rapid screening of innate immune gene expression in zebrafish using reverse transcription-multiplex ligation-dependent probe amplification. *BMC Res. Notes* **2011**, *4*, 196. [CrossRef] [PubMed]

55. Zou, Y.; Zhang, Y.; Han, L.; He, Q.; Hou, H.; Han, J.; Wang, X.; Li, C.; Cen, J.; Liu, K. Oxidative stress-mediated developmental toxicity induced by isoniazide in zebrafish embryos and larvae. *J. Appl. Toxicol.* **2017**, *37*, 842–852. [CrossRef] [PubMed]
56. Ryu, B.; Kim, C.Y.; Oh, H.; Kim, U.; Kim, J.; Jung, C.R.; Lee, B.H.; Lee, S.; Chang, S.N.; Lee, J.M.; et al. Development of an alternative zebrafish model for drug-induced intestinal toxicity. *J. Appl. Toxicol.* **2018**, *38*, 259–273. [CrossRef] [PubMed]
57. Brugman, S.; Witte, M.; Scholman, R.C.; Klein, M.R.; Boes, M.; Nieuwenhuis, E.E. T Lymphocyte–Dependent and–Independent Regulation of Cxcl8 Expression in Zebrafish Intestines. *J. Immunol.* **2014**, *192*, 484–491. [CrossRef]
58. Livak, J.K.; Schmittgen, D.T. Analysis of Relative Gene Expression Data Using Real-Time Quantitative PCR and the 2-ΔΔCT Method. *Methods* **2001**, *25*, 402–408. [CrossRef]
59. Caporaso, J.G.; Lauber, C.L.; Walters, W.A.; Berg-Lyons, D.; Lozupone, C.A.; Turnbaugh, P.J.; Fierer, N.; Knight, R. Global patterns of 16S rRNA diversity at a depth of millions of sequences per sample. *Proc. Natl. Acad. Sci. USA* **2011**, *108*, 4516–4522. [CrossRef]
60. Cassone, B.J.; Grove, H.C.; Elebute, O.; Villanueva, S.M.; LeMoine, C.M. Role of the intestinal microbiome in low-density polyethylene degradation by caterpillar larvae of the greater wax moth, *Galleria mellonella*. *Proc. Biol. Sci.* **2020**, *287*, 20200112. [CrossRef]
61. Quast, C.; Pruesse, E.; Yilmaz, P.; Gerken, J.; Schweer, T.; Yarza, P.; Peplies, J.; Glöckner, F.O. The SILVA ribosomal RNA gene database project: Improved data processing and web-based tools. *Nucleic Acids Res.* **2013**, *41*, 590–596. [CrossRef]
62. Gallo, F.; Fossi, C.; Weber, R.; Santillo, D.; Sousa, J.; Ingram, I.; Nadal, A.; Romano, D. Marine Litter Plastics and Microplastics and Their Toxic Chemicals Components: The Need for Urgent Preventive Measures. *Environ. Sci. Eur.* **2018**, *30*. [CrossRef]
63. Galloway, S.T.; Lewis, C.N. Marine Microplastics Spell Big Problems for Future Generations. *Proc. Natl. Acad. Sci. USA* **2016**, *113*, 2331–2333. [CrossRef] [PubMed]
64. Sharma, S.; Chatterjee, S. Microplastic Pollution, a Threat to Marine Ecosystem and Human Health: A Short Review. *Environ. Sci. Pollut. Res. Int.* **2017**, *24*, 21530–21547. [CrossRef] [PubMed]
65. Chen, Q.; Gundlach, M.; Shouye, Y.; Jiang, J.; Velki, M.; Yin, D.; Hollert, H. Quantitative Investigation of the Mechanisms of Microplastics and Nanoplastics toward Zebrafish Larvae Locomotor Activity. *Sci. Total. Environ.* **2017**, *584–585*, 1022–1031. [CrossRef] [PubMed]
66. Mazurais, D.; Ernande, B.; Quazuguel, P.; Severe, A.; Huelvan, C.; Madec, L.; Mouchel, O.; Soudant, P.; Robbens, J.; Huvet, A.; et al. Evaluation of the Impact of Polyethylene Microbeads Ingestion in European Sea Bass (*Dicentrarchus labrax*) Larvae. *Mar. Environ. Res.* **2015**, *112 Pt A*, 78–85. [CrossRef]
67. Hill, A.J.; Teraoka, H.; Heideman, W.; Peterson, R.E. Zebrafish as a Model Vertebrate for Investigating Chemical Toxicity. *Toxicol. Sci.* **2005**, *86*, 6–19. [CrossRef] [PubMed]
68. Qiang, L.; Cheng, J. Exposure to Microplastics Decreases Swimming Competence in Larval Zebrafish (*Danio rerio*). *Ecotoxicol. Environ. Saf.* **2019**, *176*, 226–233. [CrossRef]
69. Reniere, M.L. Reduce, Induce, Thrive: Bacterial Redox Sensing during Pathogenesis. *J. Bacteriol.* **2018**, *200*, e00128-18. [CrossRef]
70. Bonneaud, C.; Wilson, R.S.; Seebacher, F. Immune-Challenged Fish Up-Regulate Their Metabolic Scope to Support Locomotion. *PLoS ONE* **2016**, *11*, e0166028. [CrossRef]
71. Yuan, M.; Yan, C.; Ying, Y.H.; Lu, W. Behavioral and Metabolic Phenotype Indicate Personality in Zebrafish (*Danio rerio*). *Front. Physiol.* **2018**, *9*, 653. [CrossRef]
72. Lee, W.S.; Cho, H.-J.; Kim, E.; Huh, Y.H.; Kim, H.-J.; Kim, B.; Kang, T.; Lee, J.-S.; Jeong, J. Bioaccumulation of Polystyrene Nanoplastics and Their Effect on the Toxicity of Au Ions in Zebrafish Embryos. *Nanoscale* **2019**, *11*, 3200–3207. [CrossRef]
73. Novoa, B.; Figueras, A. Zebrafish: Model for the Study of Inflammation and the Innate Immune Response to Infectious Diseases. *Adv. Exp. Med. Biol.* **2012**, *946*, 253–275. [CrossRef] [PubMed]
74. Akira, S.; Hoshino, K. Myeloid Differentiation Factor 88–Dependent and –Independent Pathways in Toll-Like Receptor Signaling. *J. Infect. Dis.* **2003**, *187* (Suppl. 2), S356–S363. [CrossRef]
75. Liu, T.; Zhang, L.; Joo, D.; Sun, S.-C. NF-KB Signaling in Inflammation. *Signal Transduct. Target Ther.* **2017**, 17023. [CrossRef] [PubMed]

76. Sar, A.M.; Stockhammer, O.W.; Laan, C.; Spaink, H.P.; Bitter, W.; Meijer, A.H. MyD88 Innate Immune Function in a Zebrafish Embryo Infection Model. *Infect. Immun.* **2006**, *74*, 2436–2441. [CrossRef] [PubMed]
77. Brugman, S. The Zebrafish as a Model to Study Intestinal Inflammation. *Dev. Comp. Immunol.* **2016**, *64*, 82–92. [CrossRef] [PubMed]
78. Hanyang, L.; Liu, X.; Chen, X.; Yujia, Q.; Jiarong, F.; Jun, S.; Zhihua, R. Application of Zebrafish Models in Inflammatory Bowel Disease. *Front. Immunol.* **2017**, *8*. [CrossRef] [PubMed]
79. Wang, G.; Bonkovsky, H.L.; de Lemos, A.; Burczynski, F.J. Recent Insights into the Biological Functions of Liver Fatty Acid Binding Protein 1. *J. Lipid Res.* **2015**, *56*, 2238–2247. [CrossRef]
80. Maragakis, N.J.; Rothstein, J.D. Motor Neuron Disease: Amyotrophic Lateral Sclerosis. *Mol. Neurobiol.* **2007**, 307–319. [CrossRef]
81. Jeong, C.B.; Kang, H.M.; Lee, M.C.; Kim, D.H.; Han, J.; Hwang, S.; Souissi, S.; Lee, S.J.; Shin, K.H.; Park, H.G.; et al. Adverse Effects of Microplastics and Oxidative Stress-Induced MAPK/Nrf2 Pathway-Mediated Defense Mechanisms in the Marine Copepod *Paracyclopina nana*. *Sci. Rep.* **2017**, *7*, 1–11. [CrossRef]
82. Egerton, S.; Culloty, S.; Whooley, J.; Stanton, C.; Ross, P.R. The Gut Microbiota of Marine Fish. *Front. Microbiol.* **2018**, *9*, 873. [CrossRef]
83. Roeselers, G.; Mittge, E.K.; Stephens, Z.W.; Parichy, D.M.; Cavanaugh, C.M.; Guillemin, K.; Rawls, J.F. Evidence for a Core Gut Microbiota in the Zebrafish. *ISME J.* **2011**, *5*, 1595–1608. [CrossRef] [PubMed]
84. Stephens, W.Z.; Burns, R.A.; Stagaman, K.; Wong, S.; Rawls, J.F.; Guillemin, K.; Bohannan, B.J. The Composition of the Zebrafish Intestinal Microbial Community Varies across Development. *ISME J.* **2016**, *10*, 644–654. [CrossRef] [PubMed]
85. Phelps, D.; Brinkman, N.E.; Keely, S.P.; Annken, E.M.; Catron, T.R.; Betancourt, D.; Wood, C.E.; Espenschied, S.T.; Rawls, J.F.; Tal, T. Microbial Colonization Is Required for Normal Neurobehavioral Development in Zebrafish. *Sci. Rep.* **2017**, *7*, 11244. [CrossRef] [PubMed]
86. Harrison, J.P.; Sapp, M.; Schratzberger, M.; Osborn, M.A. Interactions Between Microorganisms and Marine Microplastics: A Call for Research. *Mar. Technol. Soc. J.* **2011**, *45*, 12–20. [CrossRef]
87. Lu, K.; Ruxia, Q.; Hao, A.; Zhang, Y. Influence of Microplastics on the Accumulation and Chronic Toxic Effects of Cadmium in Zebrafish (*Danio rerio*). *Chemosphere* **2018**, *202*, 514–520. [CrossRef]
88. Prakash, S.; Rodes, L.; Coussa-Charley, M.; Tomaro-Duchnesneau, C. Gut Microbiota: Next Frontier in Understanding Human Health and Development of Biotherapeutics. *Biologics* **2011**, *5*, 71–86. [CrossRef]
89. Vael, C.; Nelen, V.; Verhulst, L.S.; Goossens, H.; Desager, K.N. Early Intestinal Bacteroides Fragilis Colonisation and Development of Asthma. *BMC Pulm. Med.* **2008**, *8*, 19. [CrossRef]
90. Cadet, J.; Wagner, R.J. DNA Base Damage by Reactive Oxygen Species, Oxidizing Agents, and UV Radiation. *Cold Spring Harb. Perspect. Biol.* **2013**, *5*, a012559. [CrossRef]
91. Dumitrescu, L.; Popescu-Olaru, I.; Cozma, L.; Tulbă, D.; Eugen, M.H.; Ceafalan, L.C.; Gherghiceanu, M.; Popescu, B.O. Oxidative Stress and the Microbiota-Gut-Brain Axis. *Oxid. Med. Cell Longev.* **2018**, 406594. [CrossRef]
92. Johnson, L.A.; Hug, L.A. Distribution of reactive oxygen species defense mechanisms across domain bacteria. *Free Radic. Biol. Med.* **2019**, *140*, 93–102. [CrossRef]
93. Machado, M.V.; Cortez-Pinto, H. Gut Microbiota and Nonalcoholic Fatty Liver Disease. *Ann. Hepatol.* **2012**, *11*, 440–449. [CrossRef]
94. Suzuki, A.; Ito, M.; Hamaguchi, T.; Mori, H.; Takeda, Y.; Baba, R.; Watanabe, T.; Kurokawa, K.; Asakawa, S.; Hirayama, M.; et al. Quantification of Hydrogen Production by Intestinal Bacteria That Are Specifically Dysregulated in Parkinson's Disease. *PLoS ONE* **2018**, *13*, e0208313. [CrossRef] [PubMed]

Article

Occurrence of Microplastic Pollution at Oyster Reefs and Other Coastal Sites in the Mississippi Sound, USA: Impacts of Freshwater Inflows from Flooding

Austin Scircle [1], James V. Cizdziel [1,*], Louis Tisinger [2], Tarun Anumol [2] and Darren Robey [2]

1 Department of Chemistry and Biochemistry, University of Mississippi, University, MS 38677, USA; arscircl@go.olemiss.edu

2 Agilent Technologies Inc., 2850 Centerville Road, Wilmington, DE 19808, USA; louis.tisinger@agilent.com (L.T.); tarun.anumol@agilent.com (T.A.); darren.robey@agilent.com (D.R.)

* Correspondence: cizdziel@olemiss.edu

Received: 17 April 2020; Accepted: 13 May 2020; Published: 15 May 2020

Abstract: Much of the seafood that humans consume comes from estuaries and coastal areas where microplastics (MPs) accumulate, due in part to continual input and degradation of plastic litter from rivers and runoff. As filter feeders, oysters (*Crassostrea virginica*) are especially vulnerable to MP pollution. In this study, we assessed MP pollution in water at oyster reefs along the Mississippi Gulf Coast when: (1) historic flooding of the Mississippi River caused the Bonnet Carré Spillway to remain open for a record period of time causing major freshwater intrusion to the area and deleterious impacts on the species and (2) the spillway was closed, and normal salinity conditions resumed. Microplastics (~25 μm–5 mm) were isolated using a single-pot method, preparing samples in the same vessel (Mason jars) used for their collection right up until the MPs were transferred onto filters for analyses. The MPs were quantified using Nile Red fluorescence detection and identified using laser direct infrared (LDIR) analysis. Concentrations ranged from ~12 to 381 particles/L and tended to decrease at sites impacted by major freshwater intrusion. With the spillway open, average MP concentrations were positively correlated with salinity ($r = 0.87$, $p = 0.05$) for sites with three or more samples examined. However, the dilution effect on MP abundances was temporary, and oyster yields suffered from the extended periods of lower salinity. There were no significant changes in the relative distribution of MPs during freshwater intrusions; most of the MPs (>50%) were in the lower size fraction (~25–90 μm) and consisted mostly of fragments (~84%), followed by fibers (~11%) and beads (~5%). The most prevalent plastic was polyester, followed by acrylates/polyurethanes, polyamide, polypropylene, polyethylene, and polyacetal. Overall, this work provides much-needed empirical data on the abundances, morphologies, and types of MPs that oysters are exposed to in the Mississippi Sound, although how much of these MPs are ingested and their impacts on the organisms deserves further scrutiny. This paper is believed to be the first major application of LDIR to the analysis of MPs in natural waters.

Keywords: plastic pollution; oysters; Mississippi Sound; fluorescence microscopy; laser direct infrared analysis; LDIR; bulk water sampling; Bonnet Carré Spillway

1. Introduction

The occurrence of microplastics (MPs) in the aquatic environment is well documented and is the subject of increasing governmental and public attention [1–3]. Microplastics have been detected in practically all water bodies that have been studied, including the Arctic Ocean and remote mountain lakes [4,5]. Owing to their small size and pervasiveness in marine waters, MPs can pose a serious threat to certain aquatic organisms, particularly filter-feeding organisms such as mollusks and oysters [2].

Further, the majority of seafood comes from estuaries and coastal areas where MPs accumulate, due to continual input and degradation of plastic litter from rivers and runoff [1].

Several studies have shown that MPs can interfere with nutritional uptake, reproduction, and offspring performance in oysters and mussels and influence larval fish ecology [6–9], but another study has shown that while MPs are readily ingested by oyster larvae, exposure to plastic concentrations exceeding those observed in the environment resulted in no measurable effects on the development or feeding capacity of the larvae over the duration of the study [10]. In any case, MPs are largely regarded not only as an emerging threat to these susceptible species, but also as an additional stressor to marine ecosystems as a whole [11]. Oyster populations have declined due to a variety of converging factors including oil spills, anoxic and freshwater influx events, disease, invasive species, and overfishing, which have culminated in the loss of ~85% of oyster reefs globally and in the functional extinction of oyster populations in areas where oysters once flourished [12]. Given the precarious state of oyster populations, toxicological studies on the effects MPs have on oysters desperately need field studies to provide critical information on the quantity, sizes, and types of MPs found at oyster reefs. However, such empirical data are generally lacking.

There are a few papers on MPs in oysters and at oyster reefs. Li et al. [13] examined MPs in oysters along the Pearl River Estuary in China, finding 1.5 to 7.2 items per gram of tissue (wet weight), which was positively correlated with concentrations in the surrounding water. Keisling et al. [14] found an average of 0.72 MPs per individual oyster (0.18 particles/g). Rochman et al. [15] found plastic debris and fibers > 500 μm in size in 33% of oysters sold for human consumption. Most such studies rely on visual inspection, and undoubtedly many smaller particles go undetected. Thus, MPs are a potential and poorly understood threat to oyster populations and nearby human communities, like those along the Gulf Coast, where oyster farming is a vital part of the economy. Moreover, the levels of MPs in the northern Gulf of Mexico (nGoM) are among the highest reported globally, both on beaches [16] and in the water, where recent surveys along the Louisiana coast measured as much as 150,000 MPs/m^3 [17].

We previously reported on MP pollution in the Mississippi River basin using a novel single-pot method for sample preparation and Nile Red fluorescence for detection and quantitation [18]. Here, we use the same approach to, for the first time, quantify the abundances and morphology of MPs in water at four different oyster reefs and six other sites in the Mississippi Sound located in the nGoM. We also used LDIR analysis to identify the major types of plastics in a subset of the samples. LDIR uses the long-established infrared spectroscopic technique that has been updated and automated in the Agilent's 8700 system. To the best of our knowledge, this is the first major application of LDIR analyses for the determination, characterization, and identification of MPs in natural waters. Grab samples were collected seasonally for a year, with spring and summer samples taken when historic flooding of the Mississippi River resulted in major freshwater influxes into the western portion of the Mississippi Sound that devastated oyster reefs and with fall and winter samples collected under more normal salinity conditions. Thus, this work also represents a side-by-side comparison of MPs in the Mississippi Sound during a period of intense freshwater intrusion and during more stable salinity conditions.

2. Materials and Methods

2.1. Study Site

The Mississippi Sound is a 145 km sound located along the coasts of Mississippi and Alabama in the northern Gulf of Mexico. It has significant commercial and ecological importance for the area and is known for the harvesting of shellfish. However, oyster populations, particularly in the western portion of the sound, have been curtailed in recent years due to pollution (e.g., oil spills) and weather events, including hurricanes and flooding. In this study, we collected water samples from 10 sites, 4 of which were directly above oyster reefs, with the remainder spread throughout the Mississippi Sound (Figure 1). Specific sampling locations (GPS coordinates) and certain water quality measurements are provided in Table 1. Samples were collected in the spring and summer of 2019 during a period of major (historic) flooding along the Mississippi River, as well as in September 2019 and January and February 2020 during more normal salinity conditions (post-flood). During periods of flooding along the lower Mississippi River, its waters are diverted through the Bonnet Carré Spillway, preventing flooding downstream in the city of New Orleans. These floodwaters spill into Lake Ponchartrain, then Lake Borgne, and from there enter the western end of the Mississippi Sound.

Figure 1. Map showing the sampling locations (circles) in the Mississippi Sound with direction of freshwater inflows from the Bonnet Carré Spillway and the Pearl River (red arrows). Sample site numbers increase from west to east. Sampling sites 2, 4, 6, and 8 (red stars) were at oyster reefs. Red circles represent sites where samples were collected by the Mississippi Department of Environmental Quality, and blue circles sites where samples were collected by the University of Mississippi. Site 10 is off the map, about 40 km to the east in Alabama waters. The inset shows the general study location within the USA.

Table 1. Geographic coordinates for sites where water was collected along the Mississippi Gulf Coast for microplastic analyses. Sites are listed west to east, with site numbers depicted in Figure 1.

Site #	Site Name	Reef Site	GPS Coordinates		Depth (m)	Sampling Dates	
			Lat.	Lon.		Open Spillway	Closed Spillway
1	St. Joe's Pass	No	30.1068	−89.5528	3.7	July (2019)	January (2020)
2	Waveland Reef	Yes	30.2730	−89.3702	2.6	April (2019)	September (2019)
3	Bay St. Louis	No	30.3510	−89.3547	1.3	April (2019)	September (2019)
4	St. Stanislaus Reef	Yes	30.3023	−89.3272	1.9	April (2019)	September (2019)
5	TNC Bay St Louis	No	30.3451	−89.2949	1.5	April (2019)	September (2019)
6	Henderson Pt.	Yes	30.2926	−89.2711	3.0	April (2019)	September (2019)
7	Pass Christian	No	30.2850	−89.2371	3.8	July (2019)	January (2020)
8	Kittiwake Reef	Yes	30.3324	−89.1652	2.3	April (2019)	September (2019)
9	Biloxi Bay	No	30.3753	−88.8306	1.4	July (2019)	January (2020)
10	Middle Bay	No	30.3749	−88.3992	1.2	July (2019)	January (2020)

2.2. *Sampling the Waters of the Mississippi Sound*

In four separate sampling campaigns, we obtained bulk water (grab) samples from the Mississippi Sound during periods when the Bonnet Carré Spillway was open and dumping freshwater into the Sound and when it had been closed. Our research group at the University of Mississippi (UM) collected samples during April (open spillway) and September (closed spillway) 2019 from sites 2, 3, 4, 5, 6, 8, and the Mississippi Department of Environmental Quality (MDEQ) obtained samples in July 2019 (open spillway) and January 2020 (closed spillway) from sites 1, 7, 9, and 10 (Figure 1 and Table 1).

Grab samples were collected just below the surface into glass Mason (canning) jars (946 mL) and tightly capped using metal lids. MDEQ also collected samples <1 m above the bottom using a 4 L Van Dorn sampler, transferring the water to Mason jars (946 mL) immediately after collection. The samples were placed in coolers and shipped to the laboratory at UM for processing. Mason jars are available from many grocery or hardware stores. We used jars that have a 946 mL capacity, but slightly larger (1 L) jars are readily available outside of the USA. Hereafter, we report abundances of MPs/L, adjusting for differences in volume.

2.3. Sample Preparation Using the Single-Pot Method

Samples were processed at our UM microplastics-dedicated laboratory in the same jar they were collected in, using our "single-pot" method described elsewhere [18]. Briefly, the solid canning jar lids were replaced with lids that had a 57 mm diameter opening cut into them, and between that and the jar was placed a round 84 mm-diameter 200 × 600 mesh (~25 um) screen of Monel, a rust-resistant, nickel–copper alloy (Unique Wire Weaving Co. Inc. Hillside, NJ, USA). The jars were inverted, and a stream of air was applied to break the surface tension and force the water to pass through the screen. The screen was then removed and carefully rinsed back into the Mason jar with ultrapure water (purified and filtered (0.2 µm); Milli-Q, Millipore, Burlington, MA, USA). The volume of water used for rinsing was kept low (<~100 mL) to avoid diluting the reactants in the next step, but sufficient to fully rinse the mesh. We used 20 mL of 30% H_2O_2 and 20 mL of 0.05 M Fe(II) solution (Fenton's reagent) to remove natural organic debris in the samples; the method has shown to be an effective pre-treatment for micro-spectroscopic imaging and it avoids acids and prolonged high temperatures known to damage MPs [19–22]. The solution was vacuum-filtered onto a 10 µm-pore size, 25 mm polycarbonate track-etched filter (Sterlitech Corp., Kent, WA, USA). Method blanks, consisting of ultra-pure H_2O, were prepared alongside each batch of samples and were subjected to the full sample preparation scheme (filtering, digestion, and staining).

2.4. Enumeration of Microplastics Using Fluorescence Microscopy

We used Nile red dye fluorochrome and fluorescence microscopy to detect and characterize the size distribution and morphology of the MPs as detailed elsewhere [18,21]. Briefly, a few drops of Nile red dye (10 µg/mL in methanol) were placed onto the filters with MPs. The samples were allowed to dry for ~20 min in a clean laminar flow hood. The filters were transferred onto microscope slides, and a coverslip was taped above them to prevent particles from shifting when the slides were handled. The samples were analyzed within a day of staining using a Nikon Ti2 Eclipse Fluorescence Microscope. To detect and count putative MPs, we followed a procedure (delineated elsewhere [18]) which took <30 min for each sample. To summarize, the microscopy consisted of imaging the entire filter and counting fluorescing objects lacking biological features such as cellular structure or striations. The dimensions of the objects were determined using built-in measurement tools in the NIS-Elements application (Nikon Instruments Inc., Melville, NY, USA) used to collect the fluorescence microscopy images. The software gives automated measurement of the dimensions of all counted objects, including diameter, length, width, perimeter, area, and circularity. These measurements were used to sort the objects into one of three categories, i.e., fragments, fibers, or beads. Objects with a length-to-width ratio of 3 or greater were defined as fibers, while objects with 0.90 or greater circularity were defined as beads, and all else were categorized as fragments. Method blanks, which rarely exceeded 1/4 of sample counts, were subtracted from the sample results such that data reported herein are blank-subtracted.

2.5. Laser Direct Infrared (LDIR) Analysis of Microplastics

To identify the major types and size fraction of plastics in a subset of seawater samples, we used the Agilent 8700 LDIR analyzer at Agilent Technologies Application Laboratory in Wood Dale, IL, USA. Because this is the first major application of LDIR for microplastics in natural waters, we will briefly introduce the technique before discussing our specific analytical method.

At its core, the 8700 LDIR is simply an infrared imaging microscope. The key novel aspect is the light source, a proprietary Quantum Cascade Laser (QCL). The QCL is a semi-conductor-based laser in which electrons tunnel through a series of quantum wells and emit light. In a diode-based laser, electron–hole recombination across semiconductor bandgap emits photons. The photon wavelength is set by the chemistry of the materials used. In a QCL, electrons cascade (tunnel) through a series of quantum wells formed by thin layers of semiconductor material. Photon wavelength is not determined by the semiconductor materials but rather by the thickness and distribution of the semiconductor layers [23]. Thus, a QCL can be rapidly tuned through a wavelength range. In some ways, it can be compared to a monochromator approach, since individual wavelengths are used at each time; however, the QCL operates at speeds and wavelength accuracy orders of magnitude better than the older approaches.

The LDIR analyzer used in this study combines the QCL with a single-point mercury cadmium telluride (MCT) detector (thermometrically cooled) and rapid scanning optics, allowing two useful modes of action. In the first, the LDIR parks the frequency at a single wavelength and scans through the objective as it moves over the sample. At each point (pixel), the information is acquired in as little as 40 μs, allowing fast scanning of large areas, significantly faster than traditional with FTIR spectroscopes. In the second mode, the objective is parked at a single point, while the QCL sweeps through the frequency range. A full spectrum is obtained in <1 s.

The microplastics analysis presented herein utilized both modes. First, the objective rapidly scanned the sample area at the parked frequency. The analysis area varied depending on how the sample was deposited on the slide, but was typically ~500 mm^2 (range ~145 mm^2 to ~1100 mm^2). In any case, the sweep locates particles in the analysis area and obtains information that can be used to describe the size and shape of the particles. Image analysis then rapidly defines the particles and describes their major axis size, area, and shape. Once located, the LDIR then rapidly and automatically moves around the analysis area of each particle and, using the parked objective/frequency sweep mode, obtains a full spectrum in the mid-IR range (1800 cm^{-1} to 975 cm^{-1}), conducts a library search, and provides a match. The library used in this study was created from data derived from external sources to include plastics of many different kinds. The library features were created from a database published earlier [24]. For the smallest particles (<30 μm), the system may determine a need to automatically refocus to obtain an optimum spectrum. In this case, the per particle analysis time may be up to 8 s.

For LDIR analysis, we prepared seawater samples as described before (see Sample preparation using the single-pot method), except we did the final filtering using a wire mesh screen with ~25 μm openings. The screen was then submerged in 50% ethanol in a glass vial and subjected to ultrasonication for ~2 min. The screen was then removed with forceps, and the vial was sealed and shipped overnight to Agilent. There, the samples were deposited on MirrIR low-e Kevley microscope slides. The solvent was allowed to evaporate, leaving behind the MPs adhering to the slide.

The LDIR analyzer was run in trans-reflectance mode, where the system directs IR laser light through the sample (particle), and the light was then reflected back off the reflective slide through the particle as it exited. As noted, the analyzer first scans the slide to locate each particle using IR light at 1800 cm^{-1}, a frequency at which little or no absorption occurs, but the light is scattered when encountering a particle. The system utilizes image analysis techniques to determine the boundary of the particle and hence the dimensions. In addition, a full spectrum of each particle covering the range of the instrument is collected and compared to the spectral library in real time. The particle is identified based on this comparison to a spectral database built into the software. Validation of this method was achieved through the use of a blind sample consisting of a blank spiked with MPs composed of polyethylene, polyamide (nylon), polypropylene, and polystyrene, and each of these types of MPs were identified in the analyses. All data analysis and processing were done in real time using the Agilent Clarity Software.

3. Results and Discussion

We found no significant difference (t-test, $\alpha = 0.05$, unequal variances) between the samples collected just below the surface ($n = 48$) and those collected near the bottom ($n = 28$) during the same sampling event and across all sampling sites. The lack of difference by depth is not surprising, given the shallow waters (often <3 m deep) and mixing by wave action. Hereafter, we will not distinguish samples by depth.

3.1. Spatial and Temporal Trends of MPs in the Mississippi Sound

Microplastic concentrations from all sites and sampling campaigns are summarized in Table 2. The average concentrations ranged from 30 to 192 particles/L across the sites, with concentrations of individual samples ranging from 12 to 381 particles/L. It is not unusual to observe such a wide range of MP concentrations, especially when counting particles as low as 30 μm in bulk water grab samples. Indeed, the sampling method will affect the counts and distribution of the MPs measured [25]. For example, nets are typically used in investigating large areas, with results being reported in particles/m^3, whereas bulk water sampling, done here, can provide a snapshot at a given site and is generally reported in particles/L. A major drawback to sampling with a net is that it fails to capture particles smaller than the mesh opening (typically 333 μm), and these smaller particles tend to be the most abundant [26]. Conversely, the probability of capturing larger particles in grab samples is lower compared to trawling with a net.

Table 2. Concentrations of microplastics (MPs, >~25 μm–5 mm) along the Mississippi Gulf Coast during freshwater inflows in the summer and fall of 2019 and winter of 2020 [a].

Site #	Location/Name	Open Spillway				Closed Spillway			
		n	Mean (Range) (MPs/L)	SD	Salinity (ppth) [b]	n	Mean (Range) (MPs/L)	SD	Salinity (ppth) [b]
1	St. Joe's Pass	2	196 (12–381)	NA	0.39	1	309	NA	NA
2	Waveland Reef	5	73 (55–153)	50	0.32	2	174 (39–309)	NA	16.4
3	Bay St. Louis	4	49 (18–80)	25	0.67	3	50 (0–116)	60	NA
4	St. Stanislaus Reef	4	36 (18–65)	20	0.40	5	121 (34–99)	76	16.8
5	TNC Bay St. Louis	2	69 (38–100)	NA	0.54	3	39 (15–73)	30	15.3
6	Henderson Pt. Reef	3	30 (20–50)	17	3.14	5	139 (20–198)	76	17.2
7	Pass Christian	12	178 (55–328)	223	8.1	3	176 (124–202)	44	4.7
8	Kittiwake Reef	4	121 (22–196)	82	4.21	-	NA	NA	18.9
9	Biloxi Bay	8	179 (64–278)	82	13.5	3	94 (55–116)	34	9.3
10	Middle Bay	6	192 (39–326)	107	19	3	141 (41–263)	112	17

[a] Data are blank-subtracted; NA = not available. [b] ppth = part per thousand; oysters require at least 8 ppth salinity to grow (Virginia Dept. of Environmental Quality).

Using the average concentration of MPs determined throughout the study (129 ± 93 MPs/L) and the average volume of water that an adult oyster filters daily (~189 L) [27], we estimated oysters may be exposed to nearly 24,000 MPs daily (range ~5600 to ~36,000). We stress that this is only an estimate and that MP concentrations and filtering rates will vary depending on site-specific conditions, oyster species, and other factors. Moreover, whether the MPs are actually entrained in oyster tissues likely depends on their size and morphologies and whether they are ingested or just make contact with the tissues (see microplastic morphologies below). Recent studies on oyster ingestion of MPs have shown that oysters are ingesting and retaining between 0.6 and 16.5 MPs per individual [14,15,28–31]. These studies also show that oysters nearer urban centers often contained higher concentrations of MPs. Given the prevalence of commercial fishing, oil drilling, and shipping ports in the nGoM, oysters along the Gulf Coast could be accumulating a considerable amount of MPs.

Despite the inherent variability in MP concentrations, we observed a moderately positive correlation ($r = 0.62$) between the average MP concentration and salinity at each site (salinity being used as a proxy for lack of freshwater intrusion) when the spillway was open. The correlation (r-value) improved to 0.87 after removing data from sites that had only two data points (sites 1 and 5) (Figure 2). There was no such trend when the spillway was closed, suggesting a link between freshwater inflows and MP concentrations, essentially a dilution effect during high river discharges. Indeed, for the majority of the sites, MP concentrations were higher when the spillway was closed (Table 2). Further, the MP concentrations in the seawater were typically higher than what we observed for the Mississippi River and its tributaries [18]. We conclude that seawater along the Mississippi Gulf Coast has higher MP abundances (when including MPs sizes down to ~30 μm) than the river supplying new loads of MPs to the coastal area. This finding that the estuaries have higher concentrations of MPs than their riverine inputs has been observed elsewhere [27]. This is not surprising, as the river is continually flushed of plastics, the estuary serves as a sink for these plastics, and with time the plastics mechanically and photolytically degrade to smaller and smaller particles.

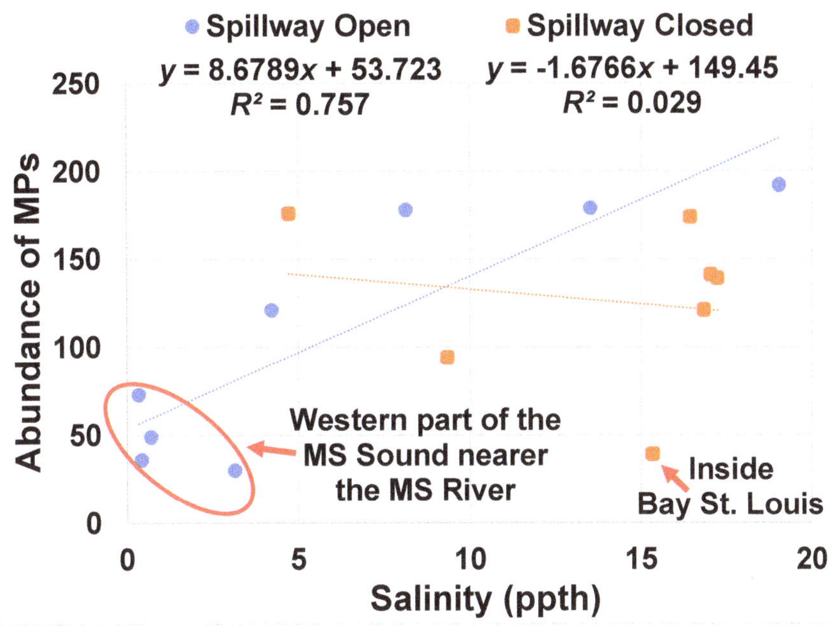

Figure 2. Salinity versus microplastic abundances in the water from sites across the Mississippi (MS) Sound, showing a moderate correlation during freshwater intrusion from flooding (spillway open). All sites were from open water areas in the Gulf, except for Bay St. Louis. Data are means of three or more measurements. ppth = parts per thousand.

Because a high degree of spatial variability is common in MP studies [32,33], especially when counting smaller size fractions like in this study, it can be difficult to observe trends, particularly where there are no major point sources. Here, except for during intense freshwater inputs in the western portion of the Sound during the spring and summer, MP concentrations did not show any distinct trend in open water areas based on longitude. However, sites 3 and 5, both located inside Bay St. Louis, tended to have lower concentrations compared to adjacent open waters of the Sound. For example, when the spillway was closed (i.e., "normal" conditions) the average concentration inside the Bay was 54 ± 18 MPs/L (n = 5; ±1 SE), while just outside the Bay, the levels were 145 ± 18 MPs/L (n = 13; ±1 SE) (see also Table 2). The reason for this is unclear and requires further study, but the circulation, mixing, and flow inside Bay St. Louis are certainly different than those of the other coastal open water sites. Additionally, the bay has its own freshwater input coming from the Wolf River, which could further contribute to the lower concentrations observed. Interestingly, the size distributions of MPs in the samples were generally similar between periods when the spillway was open and when it was closed, with the smallest size fraction, 20–90 μm, being the most common and comprising ~70% of the counted particles.

3.2. Microplastic Morphologies

The microplastic morphologies determined by fluorescence microscopy were generally consistent from site to site (Figure 3). The morphologies were similarly distributed whether the water was gathered in Bay St. Louis or at a reef in the Mississippi Sound. Fragments made up the majority of MPs counted (77–88%), followed by fibers (7–15%), and beads (1–8%). This contrasts with studies that have predominantly found anthropogenic fibers in oyster tissue [14,15,31]. However, most of these studies employed optical microscopy for detection instead of fluorescence microscopy, targeting larger microplastics. A study that used μFT-IR analyses for MP contamination in bivalves found that most of the MPs were indeed fragments [34]. However, it is also possible that fibers may be disproportionally retained because of their shape. Regardless, the connection between levels of MP pollution at reefs and concentrations of MPs ingested by oysters may not exhibit a linear relationship and requires further scrutiny.

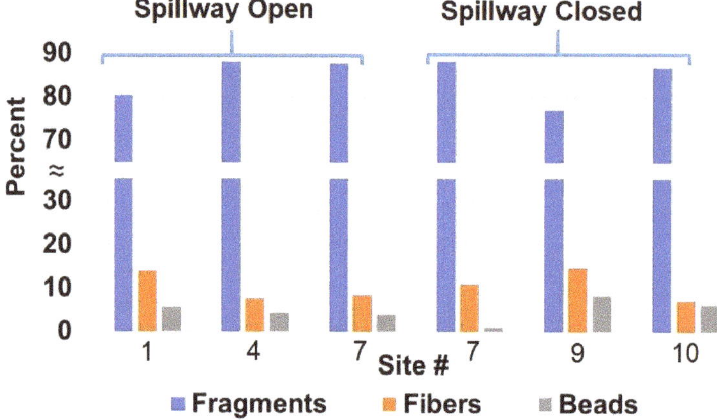

Figure 3. Morphologies of microplastics in water from select sites in the Mississippi Sound. Site number information is given in Table 1.

3.3. Identification and Quantification of Microplastics Using LDIR

LDIR analysis was applied to identify putative MPs for a selection of samples from sites 1, 7, 9, and 10 (see Figure 1 for locations and Table 3 for LDIR results). The most identified plastic throughout the samples was polyester, followed by acrylates/polyurethanes, and polyamide. Other plastics identified were polypropylene, polyethylene, polyacetal, and polytetrafluoroethylene (PTFE). Whereas polyesters and polyamides are typically used in synthetic fibers, all of the identified plastics are commonly encountered in similar studies [1,2,27,35].

Table 3. Identification of microplastics in a subset of samples from the Mississippi Sound by LDIR.

Location (Site # in Figure 1)	Spillway	MP Counts	Most Abundant Plastics Identified (% of Total)						
			Poly-Ester	Acrylates/PU	Poly-Amide	PP	PE	PA	PTFE
St. Joe's Pass (1)	closed	1154	44.3	17.2	10.1	9.0	2.5	7.7	2.3
Pass Christian (7)	open	1061	20.4	42.3	14.4	7.8	5.2	0.9	0.8
Biloxi Bay (9)	closed	383	76.2	10.3	1.9	0.9	0.7	5.3	2.8
Middle Bay (10)	open	181	31.7	14.5	7.9	8.4	2.2	24.2	3.5
Middle Bay (10)	closed	76	36.1	30.3	6.6	1.6	5.7	3.3	6.6

PU = polyurethanes; PP = polypropylene; PE = polyethylene; PA = polyacetal; PTFE = polytetrafluoroethylene.

LDIR data was also used to quantify the abundance of MPs (see MP counts in Table 3, third column) and their size distribution in the same analytical run (Figure 4). The LDIR data followed the previously discussed trend, with lower numbers of particles in the water when the spillway was open compared to when it was closed. The data also showed a general increase in the overall number of MPs at sites further to the west nearer the mouth of the Mississippi River, a major source of plastic pollution (Table 3). However, given the limited number of samples that were analyzed by LDIR, we hesitate to make a detailed comparison with fluorescence microscopy. Nevertheless, it is worth noting that, while the absolute numbers of particles detected by the two techniques were in the same magnitude (usually between a hundred and a thousand particles), LDIR tended to detect more particles. It should be emphasized that, unlike fluorescence microscopy that involves non-targeted staining of particles and assumes fluorescing particles are plastic, LDIR produces spectra of individual particles, confirming their identity. Moreover, regardless of the technique used to identify particles, some particles remain unidentified; these could be mixtures of polymers or polymers with adhering particles or biofilms, which complicate the spectra and decrease the probability of a library match. Thus, the fluorescence counts reported herein can be considered conservative.

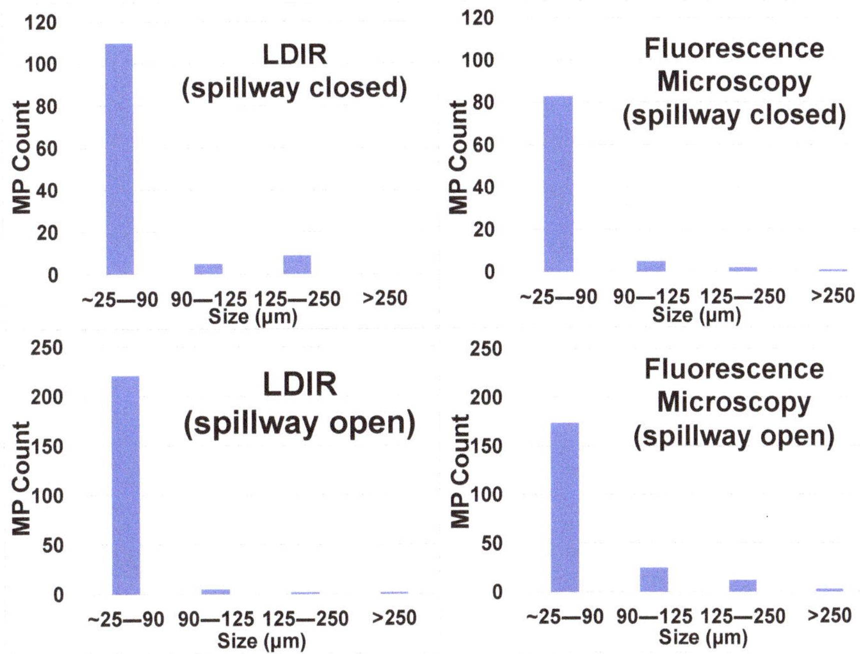

Figure 4. Particle size distribution of microplastics from 1 L of seawater collected in the Mississippi Sound and determined by Laser Direct Infrared (LDIR, **left**) and fluorescence microscopy (**right**). Note the different scale for MP counts for spillway open (**top**, indicating freshwater intrusion) and spillway closed (**bottom**, representing normal salinity conditions).

Overall, we show that the Agilent's LDIR analyzer is a powerful new automated analytical tool to rapidly detect and characterize MP pollution. A detailed comparison between LDIR and other chemical imaging techniques used for microplastic analyses will be the subject of a future report.

Author Contributions: A.S. led the sampling and measurement campaign; A.S. and J.V.C. analyzed and interpreted the data and authored the manuscript; J.V.C. conceptualized, supervised, and administered the project; T.A. and D.R. were involved in the writing and reviewing, editing, and data interpretation; L.T. conducted LDIR analyses and helped interpret the results. All authors have read and agreed to the published version of the manuscript.

Funding: This research was partially funded by the U.S. Geological Survey (USGS) under Grant/Cooperative Agreement No. (G16AP00065). The views and conclusions contained in this document are those of the authors and should not be interpreted as representing the opinions or policies of the USGS. Mention of trade names or commercial products does not constitute their endorsement by the USGS.

Acknowledgments: We are grateful to Emily Cotton and coworkers at the Mississippi Department of Environmental Quality for their sampling efforts. We also thank James Gledhill and Ann Fairly Barnett at the University of Mississippi (Department of BioMolecular Sciences) for allowing us to piggyback on their sampling trips to the oyster reefs.

Conflicts of Interest: The authors declare no conflict of interest. The funders had no role in the design of the study; in the collection, analyses, or interpretation of data; in the writing of the manuscript, or in the decision to publish the results.

References

1. Andrady, A.L. Microplastics in the marine environment. *Mar. Pollut. Bull.* **2011**, *62*, 1596–1605. [CrossRef]
2. Auta, H.; Emenike, C.; Fauziah, S. Distribution and importance of microplastics in the marine environment: A review of the sources, fate, effects, and potential solutions. *Environ. Int.* **2017**, *102*, 165–176. [CrossRef]
3. Akdogan, Z.; Guven, B. Microplastics in the environment: A critical review of current understanding and identification of future research needs. *Environ. Pollut.* **2019**, *254*, 113011. [CrossRef]
4. Lusher, A.L.; Tirelli, V.; O'Connor, I.; Officer, R. Microplastics in Arctic polar waters: The first reported values of particles in surface and sub-surface samples. *Sci. Rep.* **2015**, *5*, 789–809. [CrossRef] [PubMed]
5. Free, C.M.; Jensen, O.P.; Mason, S.A.; Eriksen, M.; Williamson, N.J.; Boldgiv, B. High-levels of microplastic pollution in a large, remote, mountain lake. *Mar. Pollut. Bull.* **2014**, *85*, 156–163. [CrossRef] [PubMed]
6. Moos, N.V.; Burkhardt-Holm, P.; Köhler, A. Uptake and effects of microplastics on cells and tissue of the blue mussel mytilus edulis L. after an experimental exposure. *Environ. Sci. Technol.* **2012**, *46*, 11327–11335. [CrossRef] [PubMed]
7. Sussarellu, R.; Suquet, M.; Thomas, Y.; Lambert, C.; Fabioux, C.; Pernet, M.E.J.; Goïc, N.L.; Quillien, V.; Mingant, C.; Epelboin, Y.; et al. Oyster reproduction is affected by exposure to polystyrene microplastics. *Proc. Natl. Acad. Sci. USA* **2016**, *113*, 2430–2435. [CrossRef]
8. Silva, P.P.G.E.; Nobre, C.R.; Resaffe, P.; Pereira, C.D.S.; Gusmão, F. Leachate from microplastics impairs larval development in brown mussels. *Water Res.* **2016**, *106*, 364–370. [CrossRef]
9. Lönnstedt, O.M.; Eklöv, P. Environmentally relevant concentrations of microplastic particles influence larval fish ecology. *Science* **2016**, *352*, 1213–1216. [CrossRef]
10. Cole, M.; Galloway, T.S. Ingestion of nanoplastics and microplastics by pacific oyster larvae. *Environ. Sci. Technol.* **2015**, *49*, 14625–14632. [CrossRef]
11. Galloway, T.S.; Lewis, C.N. Marine microplastics spell big problems for future generations. *Proc. Natl. Acad. Sci. USA* **2016**, *113*, 2331–2333. [CrossRef]
12. Beck, M.W.; Brumbaugh, R.D.; Airoldi, L.; Carranza, A.; Coen, L.D.; Crawford, C.; Defeo, O.; Edgar, G.J.; Hancock, B.; Kay, M.C.; et al. Oyster reefs at risk and recommendations for conservation, restoration, and management. *BioScience* **2011**, *61*, 107–116. [CrossRef]
13. Li, J.; Green, C.; Reynolds, A.; Shi, H.; Rotchell, J.M. Microplastics in mussels sampled from coastal waters and supermarkets in the United Kingdom. *Environ. Pollut.* **2018**, *241*, 35–44. [CrossRef] [PubMed]
14. Keisling, C.; Harris, R.D.; Blaze, J.; Coffin, J.; Byers, J.E. Low concentrations and low spatial variability of marine microplastics in oysters (*Crassostrea virginica*) in a rural Georgia estuary. *Mar. Pollut. Bull.* **2020**, *150*, 110672. [CrossRef] [PubMed]
15. Rochman, C.M.; Tahir, A.; Williams, S.L.; Baxa, D.V.; Lam, R.; Miller, J.T.; Teh, F.-C.; Werorilangi, S.; Teh, S.J. Anthropogenic debris in seafood: Plastic debris and fibers from textiles in fish and bivalves sold for human consumption. *Sci. Rep.* **2015**, *5*. [CrossRef]
16. Wessel, C.C.; Lockridge, G.R.; Battiste, D.; Cebrian, J. Abundance and characteristics of microplastics in beach sediments: Insights into microplastic accumulation in northern Gulf of Mexico estuaries. *Mar. Pollut. Bull.* **2016**, *109*, 178–183. [CrossRef]
17. Mauro, R.D.; Kupchik, M.J.; Benfield, M.C. Abundant plankton-sized microplastic particles in shelf waters of the northern Gulf of Mexico. *Environ. Pollut.* **2017**, *230*, 798–809. [CrossRef]

18. Scircle, A.; Cizdziel, J.V.; Missling, K.; Li, L.; Vianello, A. Single-Pot method for the collection and preparation of natural water for microplastic analyses: Microplastics in the mississippi river system during and after historic flooding. *Environ. Toxicol. Chem.* **2020**, *39*, 986–995. [CrossRef]
19. Tagg, A.S.; Sapp, M.; Harrison, J.P.; Ojeda, J.J. Identification and quantification of microplastics in wastewater using focal plane array-based reflectance Micro-Ft-Ir imaging. *Anal. Chem.* **2015**, *87*, 6032–6040. [CrossRef]
20. Claessens, M.; Cauwenberghe, L.V.; Vandegehuchte, M.B.; Janssen, C.R. New techniques for the detection of microplastics in sediments and field collected organisms. *Mar. Pollut. Bull.* **2013**, *70*, 227–233. [CrossRef]
21. Erni-Cassola, G.; Gibson, M.I.; Thompson, R.C.; Christie-Oleza, J.A. Lost, but found with nile red: A novel method for detecting and quantifying small microplastics (1 mm to 20 μm) in environmental samples. *Environ. Sci. Technol.* **2017**, *51*, 13641–13648. [CrossRef] [PubMed]
22. Munno, K.; Helm, P.A.; Jackson, D.A.; Rochman, C.; Sims, A. Impacts of temperature and selected chemical digestion methods on microplastic particles. *Environ. Toxicol. Chem.* **2017**, *37*, 91–98. [CrossRef] [PubMed]
23. Faist, J.; Aellen, T.; Gresch, T.; Beck, M.; Giovannini, M. Progress in Quantum Cascade Lasers. In *Mid-Infrared Coherent Sources and Applications*; NATO Science for Peace and Security Series B: Physics and Biophysics; Ebrahim-Zadeh, M., Sorokina, I.T., Eds.; Springer: Dordrecht, The Netherlands, 2008.
24. Primpke, S.; Wirth, M.; Lorenz, C.; Gerdts, G. Reference database design for the automated analysis of microplastic samples based on Fourier transform infrared (FTIR) spectroscopy. *Anal. Bioanal. Chem.* **2018**, *410*, 5131–5141. [CrossRef] [PubMed]
25. Stock, F.; Kochleus, C.; Bänsch-Baltruschat, B.; Brennholt, N.; Reifferscheid, G. Sampling techniques and preparation methods for microplastic analyses in the aquatic environment—A review. *Trends Anal. Chem.* **2019**, *113*, 84–92. [CrossRef]
26. Song, Y.K.; Hong, S.H.; Jang, M.; Kang, J.-H.; Kwon, O.Y.; Han, G.M.; Shim, W.J. Large accumulation of micro-sized synthetic polymer particles in the sea surface microlayer. *Environ. Sci. Technol.* **2014**, *48*, 9014–9021. [CrossRef]
27. Newell, R.I. Ecosystem influences of natural and cultivated populations of suspension-feeding bivalve molluscs: A review. *J. Shellfish Res.* **2004**, *23*, 51–61.
28. Li, H.-X.; Ma, L.-S.; Lin, L.; Ni, Z.-X.; Xu, X.-R.; Shi, H.-H.; Yan, Y.; Zheng, G.-M.; Rittschof, D. Microplastics in oysters Saccostrea cucullata along the Pearl River Estuary, China. *Environ. Pollut.* **2018**, *236*, 619–625. [CrossRef]
29. Waite, H.R.; Donnelly, M.J.; Walters, L.J. Quantity and types of microplastics in the organic tissues of the eastern oyster *Crassostrea virginica* and Atlantic mud crab *Panopeus herbstii* from a Florida estuary. *Mar. Pollut. Bull.* **2018**, *129*, 179–185. [CrossRef]
30. Teng, J.; Wang, Q.; Ran, W.; Wu, D.; Liu, Y.; Sun, S.; Liu, H.; Cao, R.; Zhao, J. Microplastic in cultured oysters from different coastal areas of China. *Sci. Total Environ.* **2019**, *653*, 1282–1292. [CrossRef]
31. Baechler, B.; Granek, E.; Hunter, M.; Conn, K. Microplastic concentrations in two oregon bivalve species: Spatial, temporal, and species variability. *Limnol. Oceanog.* **2019**. [CrossRef]
32. Zhao, S.; Wang, T.; Zhu, L.; Xu, P.; Wang, X.; Gao, L.; Li, D. Analysis of suspended microplastics in the Changjiang Estuary: Implications for riverine plastic load to the ocean. *Water Res.* **2019**, *161*, 560–569. [CrossRef]
33. Goldstein, M.C.; Titmus, A.J.; Ford, M. Scales of spatial heterogeneity of plastic marine debris in the northeast Pacific Ocean. *PLoS ONE* **2013**, *8*, e80020. [CrossRef] [PubMed]
34. Phuong, N.N.; Poirier, L.; Pham, Q.T.; Lagarde, F.; Zalouk-Vergnoux, A. Factors influencing the microplastic contamination of bivalves from the French Atlantic coast: Location, season and/or mode of life? *Mar. Pollut. Bull.* **2018**, *129*, 664–674. [CrossRef] [PubMed]
35. Han, M.; Niu, X.; Tang, M.; Zhang, B.T.; Wang, G.; Yue, W.; Kong, X.; Zhu, J. Distribution of microplastics in surface water of the lower Yellow River near estuary. *Sci. Total Environ.* **2020**, *707*. [CrossRef] [PubMed]

Communication

Microplastics Exposure Causes Negligible Effects on the Oxidative Response Enzymes Glutathione Reductase and Peroxidase in the Oligochaete *Tubifex tubifex*

Costanza Scopetani [1,*], Maranda Esterhuizen [1,2,3], Alessandra Cincinelli [4,5] and Stephan Pflugmacher [1,2,3]

1 Faculty of Biological and Environmental Sciences Ecosystems and Environment Research Programme, University of Helsinki, Niemenkatu 73, Lahti FI-15140, Finland; maranda.esterhuizen-londt@helsinki.fi (M.E.); stephan.pflugmacher@helsinki.fi (S.P.)
2 Korea Institute of Science & Technology (KIST Europe) Environmental Safety Group. Joint Laboratory of Applied Ecotoxicology Campus E 7.1 66123 Saarbrücken, Germany
3 Helsinki Institute of Sustainability (HELSUS), Fabianinkatu 33, 00014 Helsinki, Finland
4 Department of Chemistry "Ugo Schiff", University of Florence, Sesto Fiorentino, 50019 Florence, Italy; alessandra.cincinelli@unifi.it
5 Department of Chemistry "Ugo Schiff", University of Florence, and Consorzio Interuniversitario per lo Sviluppo dei Sistemi a Grande Interfase (CSGI), Sesto Fiorentino, 50019 Florence, Italy
* Correspondence: costanza.scopetani@helsinki.fi

Received: 23 January 2020; Accepted: 12 February 2020; Published: 15 February 2020

Abstract: Microplastics (MPs) are emerging pollutants, which are considered ubiquitous in aquatic ecosystems. The effects of MPs on aquatic biota are still poorly understood, and consequently, there is a need to understand the impacts that MPs may pose to organisms. In the present study, *Tubifex tubifex*, a freshwater oligochaete commonly used as a bioindicator of the aquatic environment, was exposed to fluorescent polyethylene microspheres (up to 10 μm in size) to test whether the oxidative stress status was affected. The mortality rate of *T. tubifex*, as well as the activities of the oxidative stress status biomarker enzymes glutathione reductase and peroxidase, were assessed. In terms of oxidative stress, no significant differences between the exposure organisms and the corresponding controls were detected. Even though the data suggest that polyethylene MPs and the selected concentrations did not pose a critical risk to *T. tubifex*, the previously reported tolerance of *T. tubifex* to environmental pollution should be taken into account and thus MPs as aquatic pollutants could still represent a threat to more sensitive oligochetes.

Keywords: microplastic; polyethylene; *Tubifex tubifex*; aquatic oligochetes; mortality; oxidative stress; glutathione reductase; peroxidase; microplastic exposure; freshwater environments

---

1. Introduction

Plastic pollution is one of the primary environmental concerns we are facing today [1]. Microplastics (MPs), typically referred to as pieces smaller than 5 mm in any dimension [2], have been found on beaches, in oceans, seas, rivers, and lakes [2–8]. One of the biggest concerns regarding MP pollution is that marine and freshwater biota can mistake MP particles for food. MPs can be ingested by benthic and pelagic organisms belonging to different trophic levels [9], including mussels [10], lugworms [11], amphipods [12], zooplankton [13], and fish [14]. Some studies showed that with ingestion possible internal damages and blockages [9,15] may occur. Ingested MPs can act as vectors for transferring chemicals, additives, and other persistent organic compounds (such as Polybrominated diphenyl ethers

(PBDEs) and Polychlorinated biphenyls (PCBs)) adsorbed from surrounding waters to biota [16–19]. However, the effects of MPs on exposed biota have not been extensively investigated, and the physiological effects remain poorly understood. Thus, there is a current need to gather data to deepen our understanding of their impacts. A few studies have demonstrated that MPs could induce oxidative stress in the organisms able to ingest them [20,21]. Browne et al. [11] studied the effect of MPs on the oxidative status of lugworms (*Arenicola marina*) demonstrating that animals exposed to polyvinyl chloride (PVC) microparticles were more susceptible to oxidative damage by up to 30%. Jeong et al. [20] showed that in response to microplastic-induced reactive oxygen species (ROS), antioxidant enzymes such as glutathione reductase (GR), glutathione peroxidase (GPx), glutathione-S-transferase (GST), and superoxide dismutase (SOD) were activated in the monogonont rotifer (*Brachionus koreanus*). The authors exposed the monogonont rotifer to three different polystyrene MP sizes (0.05, 0.5, and 6 μm) and observed that the toxicity was size-dependent, and the smaller the particles, the more toxic. More recently, Lu et al. [21] showed that the freshwater fish *Danio rerio* (zebrafish) exhibited a higher oxidative stress status after seven days of exposure to polystyrene MP, revealed by increased SOD and catalase (CAT) activities. Deng et al. [22] tested the effect of fluorescent polystyrene MP in mice, revealing increased activities for SOD and GPx and decreased CAT activity, signifying the potential health risk MPs represent to biota. Nevertheless, only a few of these studies report on the physiological effects of MP exposure in biota from aquatic environments, where MP abundance is most often reported [3–8].

The present study, therefore, investigated the toxic effects of MPs in *Tubifex tubifex*, which inhabits sediments of lakes and rivers. *T. tubifex* (also referred to as sludge or sewerage worm) lives in the uppermost sediment layers of these freshwater systems usually feeding in the sediment fraction smaller than 63 μm [23]. *T. tubifex* plays a key role in the decomposition of organic matter and bioturbation [24], dwelling in the sediment with the anterior part and keeping the tail outside, undulating it in the water to enable cutaneous respiration [6]. It is considered to be a model organism to perform sediment toxicity experiments [25,26] due to its high pollution tolerance [27] and is thus often one of the last species to disappear from a polluted habitat [24,28]. *Tubifex* worms are known to accumulate heavy metals [29–31] and organic pollutants [32] and are able to ingest MPs under natural conditions [33].

We, therefore, selected *T. tubifex* as a test subject to investigate the possible effects of MP pollution it may be exposed to in its environment. The aim of the present study was to understand whether exposure to MPs could affect the oxidative stress status of *T. tubifex*. *Tubifex* worms were exposed to fluorescent polyethylene (PE) MP particles (10 μm diameter) to evaluate the survival of the worms and the potential oxidative stress induced by MPs. PE was chosen because it is one of the most abundant polymers identified among samples collected from aqueous environments [7]. The size was selected within the size range of particles the worms could consume.

The enzyme activity of the oxidative stress biomarker enzymes GR and peroxidase (POD) were measured to check for alterations. Both enzymes, GR and POD, are known to be able to counteract the damages caused by ROS species to cells [34–37] and are thus induced in response to an increased oxidative stress status. GR plays a vital role in the antioxidant system [37,38], acting as a reductant in oxidation-reduction processes catalyzing the reduction of glutathione disulfide (GSSG) to glutathione (GSH) using NADPH as a cofactor. POD is known for playing a pivotal role in preventing H_2O_2 causing damage to DNA, proteins, and cell membranes [34,37] and has often been used in monitoring stress induced by contaminants [35,36].

2. Materials and Methods

T. tubifex was cultured in 1 L beakers in a synthetic medium at a constant temperature of 20 ± 1 °C and low light (18 μmol photons/m^{2}s) with a photoperiod of 16 h light to 8 h dark (permanent continuous culture at the University of Helsinki) for several weeks before experimentation. The synthetic medium (artificial freshwater) was exchanged every three to five days, and it consisted of de-ionized water, $CaCl_2$ [240 μg/L], KCl [6 μg/L], $MgSO_4.7H_2O$ [123 μg/L], and $NaHCO_3$ [55 μg/L]. The *T. tubifex* worms were fed with dry fish food (Sera, Mikropan, analytical constituents: crude protein 47.6%, crude fat 8.7%, crude fiber 3.3%, moisture 6.0%, crude ash 11.6%) daily.

2.1. Exposure to Microplastics

To establish the mortality of the worms in response to MP exposure, the *T. tubifex* worms were exposed to polyethylene MP particles (up to 10 μm in diameter) in four treatments; i.e., (1) *T. tubifex* in artificial freshwater containing 2 g/L MP (w/v) without sediment, (2) *T. tubifex* in artificial freshwater and sediment with the sediment containing 2 mg/g MP (w/w), (3) *T. tubifex* in artificial freshwater and sediment with the media spiked with 2 g/L MP (w/v), and (4) *T. tubifex* in artificial freshwater and sediment with both the media (2 g/L w/v MP) and the sediment (2 mg/g w/w MP) containing MP. The surviving worms were counted after 24 h, 48 h, and 120 h. For each experiment (mortality and enzyme assays, for all treatments), negative controls without MPs were conducted in replicates of five in parallel.

In order to examine the effects on the oxidative stress status, *T. tubifex* worms in sediment and artificial freshwater were exposed to the polyethylene MP particles (up to 10 μm in diameter) for 24 h in two different scenarios, i.e., MP-contaminated soil (2 mg/g w/w MP) vs. MP-contaminated media (2 g/L w/v MP). After 24 h, the treated and control worms were removed from the glass beakers and washed three times each with artificial freshwater, and the surviving worms counted.

In each experimental setup, five replicates of six adult *T. tubifex* worms per replicate were used. Each independent replicate was set up in a glass beaker containing either contaminated non-sterile sediments from Vesijärvi Lake (Lahti, Finland) or contaminated artificial freshwater or both as specified above.

2.2. Oxidative Stress Status: Enzyme Assays

After rinsing the worms with artificial freshwater, the enzymes were extracted using a shortened protocol modified from Pflugmacher [39]. *T. tubifex* worms were homogenized in 20 mM sodium phosphate buffer (pH 7.0) and stirred for 20 min followed by centrifugation at 9.000× *g* for 20 min. The supernatant (S-9 fraction) was collected and the enzyme activities were assayed immediately.

The activity of peroxidase (POD, EC 1.11.1.7) was analyzed spectrophotometrically using guaiacol as substrate [40], which is oxidized in the presence of hydrogen peroxide (H_2O_2) to octahydrotetraguaiacol. Changes in color as absorbance were measured at 436 nm over 3 min at 30 °C.

Glutathione reductase (GR, EC 1.8.1.7) activity was analyzed according to Schaedle and Bassham [41] following the glutathione disulfide-dependent NADPH oxidation at 340 nm for 3 min.

Each spectrophotometric enzyme assay was performed in triplicate per independent replicate (15 readings) using an Infinite 200 Pro plate reader (Tecan). The enzyme activity was normalized against protein content, which was determined according to Bradford [42]. Enzymatic activities are reported in nkat/mg protein, where 1 kat is the conversion rate of 1 mol of substrate per sec.

During all steps of the laboratory work, the possibility of self-contamination was taken into consideration and therefore fleece clothing or other personal items which could serve as a source of MP contamination were avoided [43].

2.3. Data Analysis

Statistical analysis was performed using IBM SPSS Statistics version 25 (2017). All data were checked for normality and homogeneity by Shapiro-Wilks and Levene's test. A normal distribution of the data was analyzed using one-way analysis of variance (ANOVA) followed by Tukey's test. For not normally distributed data, the Kruskal-Wallis test was employed to identify differences amongst the treatments, followed by the Mann-Whitney test if necessary. To assess whether survival/mortality changed with time, a repeated-measures ANOVA was performed. The results were considered significant at an alpha value of 0.05. All values are reported as mean ± standard error.

3. Results

3.1. Mortality

Contamination of the artificial water, sediment, or both water and sediment with MP (Figure 1) did not significantly affect the survival of *T. tubifex* in comparison to controls with time ($p > 0.05$). With contaminated artificial water and non-contaminated sediment, 95% ± 5% of the worms survived and with both sediment and water contaminated 90% ± 10% survived after five days. For all other treatments, 100% survived.

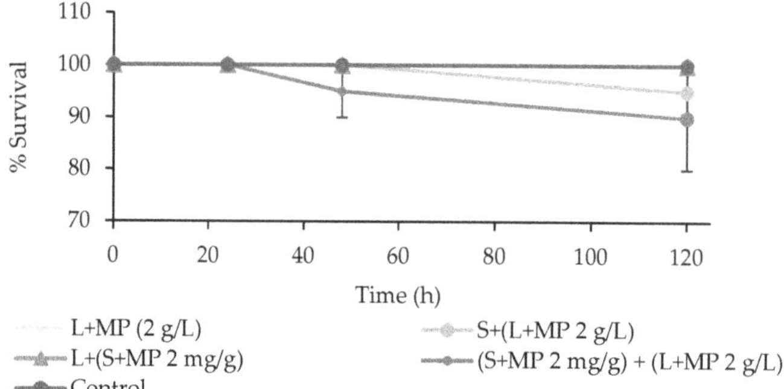

Figure 1. *Tubifex tubifex* mortality after exposure to microplastics (MPs) in media (L + MP (2 g/L)) only, in sediment and media with the media contaminated with MPs (S + (L + MP 2g/L), in sediment and media with the sediment contaminated (L + (S + MP 2 mg/g)), and sediment and media both contaminated with MPs ((S + MP 2 mg/g) + (L + MP 2 g/L)).

3.2. Oxidative Stress Status

Exposure of *T. tubifex* to fluorescence PE microspheres did not result in significant ($p > 0.05$) changes in the GR activity in comparison to the controls (Figure 2a). The mean GR activities in the worms with MP contaminated water (S + (L + MP)) and contaminated sediment (L + (S + MP)) were 0.005 ± 0.001 and 0.006 ± 0.001 nkat/mg protein, respectively, compared to the control GR activity of 0.005 ± 0.001 nkat/mg protein.

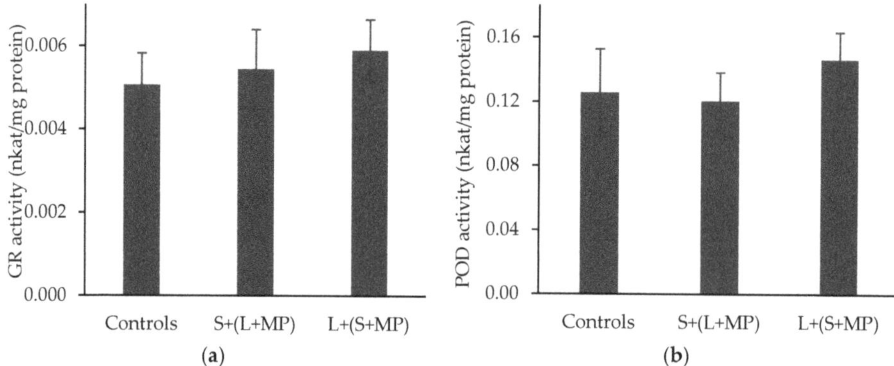

Figure 2. Peroxidase (**a**) and glutathione reductase (**b**) activities in *T. tubifex* exposed to fluorescent polyethylene (PE) microspheres through the media (S + (L + MP)) and the sediment (L + (S + MP). Values are expressed as mean enzyme activity ± standard error (n = 5). * denotes significance compared to the control ($p < 0.05$).

Peroxidase activity with exposure of *T. tubifex* to fluorescence PE microspheres, as observed for GR, did not result in significant ($p > 0.05$) changes in POD activity in comparison to controls (Figure 2b). Mean POD activities with the water (S + (L + MP)) and sediment (L + (S + MP)) contaminated respectively, were 0.120 ± 0.002 nkat/mg protein and 0.146 ± 0.002 nkat/mg protein, respectively, and the control POD activity was 0.130 ± 0.003 nkat/mg protein.

4. Discussion

Oxidative stress has been investigated intensively in many organisms [44–46] and it has been shown to cause damages to DNA, lipids, and proteins, potentially leading to an alteration in vital functions [47–50]. However, studies regarding oxidative stress and the alterations in antioxidant enzyme activities due to exposure to MPs are still limited [20,21].

In this study, *T. tubifex* was exposed to fluorescent PE microspheres to evaluate the potential damages caused by MPs as percentage mortality or survival and on a physiological level as induced oxidative stress. Mortality percentage did not show any difference between treatment and control samples. These results are in accordance with Redondo-Hasselerharm et al. [51], where no effects on the mortality rate of five freshwater benthic macroinvertebrates (including a *Tubifex sp.*) were found after exposure to polystyrene MPs for 28 days.

No significant differences were evident in terms of GR activity of *T. tubifex* exposed to MP via contaminated water and sediment. Similarly, POD activity did not differ significantly in comparison to controls in both of the experiments. These data show that the selected MP concentrations are not particularly critical for *T. tubifex* and they do not cause oxidative stress to the worms. Nevertheless, our results do not imply that MPs are not representing a threat to the biodiversity of aquatic environments. Contrary to our findings, Jeong at al., [20] and Lu et al., [21] showed that exposure to PS MPs caused oxidative stress in the monogonont rotifer (*Brachionus koreanus*) and Zebrafish (*Danio rerio*), respectively. However, we have to take into consideration the different targeted organisms and the type and size of MPs used in the studies. Either the chosen test subjects may have been more susceptible than *T. tubifex* to MPs exposure, or the selection of a different polymer could explain the discrepancy between the results. Even though exposure to PE MPs seems not to increase the production of ROS in *T. tubifex*, it does not mean that MP occurrence is not able to negatively affect freshwater environments. *T. tubifex* is considered to be a pollution-tolerant species [24,28], and the effects of MPs could pose a major risk to more sensitive organisms [11,20,21]. Furthermore, the concentration of MPs is expected to increase worldwide [52] representing a possible threat to the biodiversity of marine and freshwater environments. Further research is required to understand the effects in more sensitive organisms. In this study, no estimate on how many PE MPs were ingested has been made; a follow-up study should be performed repeating the tests changing the level of MP exposure and performing fluorescence microscopy analyses to examine the MP ingestion rate.

Author Contributions: Conventionalization and experimental design: S.P., C.S., M.E.; Analysis: C.S., M.E.; Manuscript preparation: C.S., M.E., A.C. All authors have read and agreed to the published version of the manuscript.

Funding: Open access funding provided by University of Helsinki including Helsinki University Central Hospital.

Conflicts of Interest: The authors declare no conflict of interest.

References

1. Brandon, J.A.; Jones, W.; Ohman, M.D. Multidecadal increase in plastic particles in coastal ocean sediments. *Sci. Adv.* **2019**, *5*. [CrossRef] [PubMed]
2. Martellini, T.; Guerranti, C.; Scopetani, C.; Ugolini, A.; Chelazzi, D.; Cincinelli, A. A snapshot of microplastics in the coastal areas of the Mediterranean Sea. *Trends Anal. Chem.* **2018**, *109*, 173–179. [CrossRef]
3. Cole, M.; Lindeque, P.; Halsband, C.; Galloway, T.S. Microplastics as contaminants in the marine environment: A review. *Mar. Pollut. Bull.* **2011**, *62*, 2588–2597. [CrossRef]

4. Cincinelli, A.; Martellini, T.; Guerranti, C.; Scopetani, C.; Chelazzi, D. A potpourri of microplastics in the sea surface and water column of the Mediterranean Sea. *Trends Anal. Chem.* **2018**, *110*, 321–326. [CrossRef]
5. Cincinelli, A.; Scopetani, C.; Chelazzi, D.; Lombardini, E.; Martellini, T.; Katsoyiannis, A.; Fossi, M.C.; Corsolini, S. Microplastic in the surface waters of the Ross Sea (Antarctica): Occurrence, distribution and characterization by FTIR. *Chemosphere* **2017**, *175*, 391–400. [CrossRef]
6. Liu, T.; Diao, J.; Di, S.; Zhou, Z. Bioaccumulation of isocarbophos enantiomers from laboratory-contaminated aquatic environment by tubificid worms. *Chemosphere* **2015**, *124*, 77–82. [CrossRef]
7. Klein, S.; Dimzon, I.K.; Eubeler, J.; Knepper, T.P. Analysis, occurrence, and degradation of microplastics in the aqueous environment. In *Freshwater Microplastics*; Wagner, M., Lambert, S., Eds.; Springer: Heidelberg, Germany, 2018; Volume 58, pp. 51–67.
8. Scopetani, C.; Chelazzi, D.; Cincinelli, A.; Esterhuizen-Londt, M. Assessment of microplastic pollution: Occurrence and characterisation in Vesijärvi lake and Pikku Vesijärvi pond, Finland. *Environ. Monit. Assess* **2019**, *191*, 652. [CrossRef]
9. Eerkes-Medrano, D.; Thompson, R.C.; Aldridge, D.C. Microplastics in freshwater systems: A review of the emerging threats, identification of knowledge gaps and prioritisation of research needs. *Water Res.* **2015**, *75*, 63–82. [CrossRef]
10. Farrell, P.; Nelson, K. Trophic level transfer of microplastic: *Mytilus edulis* (L.) to *Carcinus maenas* (L.). *Environ. Pollut* **2013**, *177*, 1–3. [CrossRef] [PubMed]
11. Browne, M.A.; Niven, S.J.; Galloway, T.S.; Rowland, S.J.; Thompson, R.C. Microplastic moves pollutants and additives to worms, reducing functions linked to health and biodiversity. *Curr. Biol.* **2013**, *23*, 2388–2392. [CrossRef] [PubMed]
12. Ugolini, A.; Ungherese, G.; Ciofini, M.; Lapucci, A.; Camaiti, M. Microplastic debris in sandhoppers. *Estuar. Coast. Shelf Sci.* **2013**, *129*, 19–22. [CrossRef]
13. Cole, M.; Lindeque, P.; Fileman, E.; Halsband, C.; Goodhead, R.; Moger, J.; Galloway, T.S. Microplastic ingestion by zooplankton. *Environ. Sci. Technol.* **2013**, *47*, 6646–6655. [CrossRef] [PubMed]
14. Lusher, A.L.; McHugh, M.; Thompson, R.C. Occurrence of microplastics in the gastrointestinal track of pelagic and demersal fish from the English Channel. *Mar. Pollut. Bull.* **2013**, *67*, 94–99. [CrossRef] [PubMed]
15. Wright, S.L.; Thompson, R.C.; Galloway, T.S. The physical impacts of microplastics on marine organisms: A review. *Environ. Pollut.* **2013**, *178*, 483–492. [CrossRef] [PubMed]
16. Mato, Y.; Isobe, T.; Takada, H.; Kanehiro, H.; Ohtake, C.; Kaminuma, T. Plastic resin pellets as a transport medium for toxic chemicals in the marine environment. *Environ. Sci. Technol* **2001**, *35*, 318–324. [CrossRef]
17. Ogata, Y.; Takada, H.; Mizukawa, K.; Hirai, H.; Iwasa, S.; Endo, S.; Mato, Y.; Saha, M.; Okuda, K.; Nakashima, A.; et al. International pellet watch: Global monitoring of Persistent Organic Pollutants (POPs) in coastal waters. 1. Initial phase data on PCBs, DDTs, and HCHs. *Mar. Pollut. Bull* **2009**, *58*, 1437–1446. [CrossRef] [PubMed]
18. Lithner, D.; Larsson, A.; Dave, G. Environmental and health hazard ranking and assessment of plastic polymers based on chemical composition. *Sci. Total Environ.* **2011**, *409*, 3309–3324. [CrossRef]
19. Scopetani, C.; Cincinelli, A.; Martellini, T.; Lombardini, E.; Ciofini, A.; Fortunati, A.; Pasquali, V.; Ciattini, S.; Ugolini, A. Ingested microplastic as a two-way transporter for PBDEs in Talitrus saltator. *Environ. Res.* **2018**, *167*, 411–417. [CrossRef]
20. Jeong, C.B.; Won, E.J.; Kang, H.M.; Lee, M.C.; Hwang, D.S.; Hwang, U.K.; Zhou, B.; Souissi, S.; Lee, S.J.; Lee, J.S. Microplastic size-dependent toxicity, oxidative stress induction, and p-JNK and p-p38 activation in the Monogonont rotifer (*Brachionus koreanus*). *Environ. Sci. Technol.* **2016**, *50*, 8849–8857. [CrossRef]
21. Lu, Y.; Zhang, Y.; Deng, Y.; Jiang, W.; Zhao, Y.; Geng, J.; Ding, L.; Ren, H. Uptake and accumulation of polystyrene microplastics in zebra fish (*Danio rerio*) and toxic effects in liver. *Environ. Sci. Technol.* **2016**, *50*, 4054–4060. [CrossRef]
22. Deng, Y.; Zhang, Y.; Lemos, B.; Ren, H. Tissue accumulation of microplastics in mice and biomarker responses suggest widespread health risks of exposure. *Sci. Rep.* **2017**, *7*, 46687. [CrossRef] [PubMed]
23. Rodriguez, P.; Martinez-Madrid, M.; Arrate, J.A.; Navarro, E. Selective feeding by the aquatic oligochaete *Tubifex tubifex* (Tubificidae, Clitellata). *Hydrobiologia* **2001**, *463*, 133–140. [CrossRef]
24. Mosleh, Y.Y.; Paris-Palacios, S.; Biagianti-Risbourg, S. Metallothioneins induction and antioxidative response in aquatic worms *Tubifex tubifex* (Oligochaeta, tubificiadae) exposed to copper. *Chemosphere* **2005**, *64*, 121–128. [CrossRef] [PubMed]

25. Chapman, P.M. Utility and relevance of aquatic oligochaetes in ecological risk assessment. In *Aquatic Oligochaete Biology VIII*; Rodriguez, P., Verdonschot, P.F.M., Eds.; Springer: Dordrecht, The Netherlands, 2001; Volume 158, pp. 149–169.
26. Chapman, K.K.; Benton, M.J.; Brinkhurst, R.O.; Scheuerman, P.R. Use of the aquatic oligochaetes *Lumbriculus variegatus* and *Tubifex tubifex* for assessing the toxicity of copper and cadmium in a spiked-artificial-sediment toxicity test. *Environ. Toxicol.* **1999**, *14*, 271–278. [CrossRef]
27. Wiederholm, T.; Dave, G. Toxicity of metal polluted sediments to *Daphnia magna* and *Tubifex tubifex*. *Hydrobiologia* **1989**, *176–177*, 411–417. [CrossRef]
28. Lucan-Bouché, M.L.; Biagianti-Risbourg, S.; Arsac, F.; Vernet, G. An original decontamination process developed by the aquatic oligochaete *Tubifex tubifex* exposed to copper and lead. *Aquat. Toxicol.* **1999**, *45*, 9–17. [CrossRef]
29. Bouché, M.L.; Habets, F.; Biagianti-Risbourg, S.; Vernet, G. Toxic effects and bioaccumulation of cadmium in the aquatic oligochaete *Tubifex tubifex*. *Ecotoxicol. Environ. Saf.* **2000**, *46*, 246–251. [CrossRef]
30. Hare, L.; Tessier, A.; Warren, L. Cadmium accumulation by invertebrates living at the sediment-water interface. *Environ. Toxicol. Chem.* **2001**, *20*, 880–889.
31. Méndez-Fernández, L.; Martínez-Madrid, M.; Rodriguez, P. Toxicity and critical body residues of Cd, Cu and Cr in the aquatic oligochaete *Tubifex tubifex* (Müller) based on lethal and sublethal effects. *Ecotoxicology* **2013**, *22*, 1445–1460. [CrossRef]
32. Egeler, P.; Römbke, J.; Meller, M.; Knacker, T.; Franke, C.; Studinger, G.; Nagel, R. Bioaccumulation of lindane and hexachlorobenzene by tubificid sludgeworms (Oligochaeta) under standardised laboratory conditions. *Chemosphere* **1997**, *35*, 835–852. [CrossRef]
33. Hurley, R.R.; Woodward, J.C.; Rothwell, J.J. Ingestion of microplastics by freshwater tubifex worms. *Environ. Sci. Technol.* **2017**, *51*, 12844–12851. [CrossRef] [PubMed]
34. Di Giulio, R.T.; Washburn, P.C.; Wenning, R.J.; Winston, G.W.; Jewell, C.S. Biochemical response in aquatic animals: A review of determinants of oxidative stress. *Environ. Toxicol. Chem.* **1989**, *8*, 1103–1123. [CrossRef]
35. Scalet, M.; Federico, R.; Guido, M.; Manes, F. Peroxidase activity and polyamine changes in response to ozone and simulated acid rain in Aleppo pine needles. *Environ. Exp. Bot.* **1995**, *35*, 417–425. [CrossRef]
36. Mitrovic, S.M.; Pflugmacher, S.; James, K.J.; Furey, A. Anatoxin-a elicits an increase in peroxidase and glutathione S-transferase activity in aquatic plants. *Aquat. Toxicol.* **2004**, *68*, 185–192. [CrossRef] [PubMed]
37. Monferran, M.V.; Wunderlin, D.A.; Nimptsch, J.; Pflugmacher, S. Biotransformation and antioxidant response in *Ceratophyllum demersum* experimentally exposed to 1,2- and 1,4-dichlorobenzene. *Chemosphere* **2007**, *68*, 2073–2079. [CrossRef]
38. Mishra, S.; Srivastava, S.; Tripathi, R.D.; Kumar, R.; Seth, C.S.; Grupta, D.K. Lead detoxification by coontail (*Ceratophyllum demersum* L.) involves induction of phytochelatins and antioxidant system in response to its accumulation. *Chemosphere* **2006**, *65*, 1027–1039. [CrossRef]
39. Pflugmacher, S. Promotion of oxidative stress in the aquatic macrophyte *Ceratophyllum demersum* during biotransformation of the cyanobacterial toxin microcystin-LR. *Aquat. Toxicol.* **2004**, *70*, 169–178. [CrossRef]
40. Drotar, A.; Phelps, P.; Fall, R. Evidence for glutathione peroxidase activities in cultured plant cells. *Plant Sci.* **1985**, *42*, 35–40. [CrossRef]
41. Schaedle, M.; Bassham, J.A. Chloroplast glutathione reductase. *Plant Physiol.* **1977**, *59*, 1011–1012. [CrossRef]
42. Bradford, M.M. A rapid and sensitive method for the quantification of microgram quantities of proteins utilising the principal of protein-dye binding. *Anal. Biochem.* **1976**, *72*, 248–254. [CrossRef]
43. Scopetani, C.; Esterhuizen-Londt, M.; Chelazzi, D.; Cincinelli, A.; Setälä, H.; Pflugmacher, S. Self-contamination from clothing in microplastics research. *Ecotoxicol. Environ. Saf.* **2020**, *189*, 110036. [CrossRef] [PubMed]
44. Cochón, A.C.; Della Penna, A.B.; Kristoff, G.; Piol, M.N.; San Martín de Viale, L.C.; Verrengia Guerrero, N.R. Differential effects of paraquat on oxidative stress parameters and polyamine levels in two freshwater invertebrates. *Ecotoxicol. Environ. Saf.* **2007**, *68*, 286–292. [CrossRef] [PubMed]
45. Antunes, S.C.; Freitas, R.; Figueira, E.; Gonçalves, F.; Nunes, B. Biochemical effects of acetaminophen in aquatic species: Edible clams *Venerupis decussate* and *Venerupis philippinarum*. *Environ. Sci. Pollut. Res.* **2013**, *20*, 6658–6666. [CrossRef] [PubMed]

46. Esterhuizen-Londt, M.; Schwartz, K.; Pflugmacher, S. Using aquatic fungi for pharmaceutical bioremediation: Uptake of acetaminophen by *Mucor hiemalis* does not result in an enzymatic oxidative stress response. *Fungal Biol.* **2016**, *120*, 1249–1257. [CrossRef]
47. Rikans, L.E.; Hornbrook, K.R. Lipid peroxidation, antioxidant protection and aging. *Biochim. Biophys. Acta* **1997**, *1362*, 116–127. [CrossRef]
48. Halliwell, B.; Gutteridge, J.M.C. *Free Radicals in Biology and Medicine*, 5th ed.; Oxford University Press: Oxford, UK, 1999; p. 905.
49. Barata, C.; Varo, I.; Navarro, J.C.; Arun, S.; Port, C. Antioxidant enzyme activities and lipid peroxidation in the freshwater cladoceran *Daphnia magna* exposed to redox cycling compounds. *Comp. Biochem. Physiol. C Toxicol. Pharmacol.* **2005**, *140*, 175–186. [CrossRef]
50. Esterhuizen-Londt, M.; Wiegand, C.; Downing, T.G. ß-N-Methylamino-L-alanine (BMAA) uptake by the animal model, *Daphnia magna* and consequent oxidative stress. *Toxicon* **2015**, *100*, 20–26. [CrossRef]
51. Redondo-Hasselerharm, P.E.; Falahudin, D.; Peeters, E.T.H.M.; Koelmans, A.A. Microplastic effect thresholds for freshwater benthic macroinvertebrates. *Environ. Sci. Technol.* **2018**, *52*, 2278–2286. [CrossRef]
52. Barboza, L.G.A.; Gimenez, B.C.G. Microplastics in the marine environment: Current trends and future perspectives. *Mar. Pollut. Bull.* **2015**, *97*, 5–12. [CrossRef]

Article

Enchytraeus crypticus Avoid Soil Spiked with Microplastic

Stephan Pflugmacher [1,2,3,*], Johanna H. Huttunen [1,2], Marya-Anne von Wolff [2,4], Olli-Pekka Penttinen [1,2], Yong Jun Kim [2], Sanghun Kim [5], Simon M. Mitrovic [6] and Maranda Esterhuizen-Londt [1,2,3]

1 Aquatic Ecotoxicology in an Urban Environment, Ecosystems and Environment Research Programme, Faculty of Biological and Environmental Sciences, University of Helsinki, Niemenkatu 73, 15140 Lahti, Finland; johanna.huttunen@helsinki.fi (J.H.H.); olli-pekka.penttinen@helsinki.fi (O.-P.P.); maranda.esterhuizen-londt@helsinki.fi (M.E.-L.)

2 Joint Laboratory of Applied Ecotoxicology, Environmental Safety Group, Korea Institute of Science and Technology Europe (KIST Europe) Forschungsgesellschaft mbH, Universität des Saarlandes Campus E7 1, 66123 Saarbrücken, Germany; vonwolff@tu-berlin.de (A.-M.v.W.); youngjunkim@kist-europe.de (Y.J.K.)

3 University of Helsinki, Helsinki Institute of Sustainability Science (HELSUS), Fabianinkatu 33, 00014 Helsinki, Finland

4 Department of Civil Engineering, Group of Building Materials and Construction Chemistry, Technical University of Berlin, Gustav-Meyer-Allee 25, 13355 Berlin, Germany

5 Department of Pharmaceutical Science and Technology, Centre for Chemical Safety Research, Kyungsung University, 309, Suyeong-ro, Nam-gu, Busan 48434, Korea; fatherofdamin@ks.ac.kr

6 School of Life Sciences, University of Technology Sydney, Ultimo, NSW 2007, Australia; simon.mitrovic@uts.edu.au

* Correspondence: stephan.pflugmacher@helsinki.fi; Tel.: +358-50-316-7329

Received: 23 January 2020; Accepted: 7 February 2020; Published: 10 February 2020

Abstract: Microplastics (MPs) of varying sizes are widespread pollutants in our environment. The general opinion is that the smaller the size, the more dangerous the MPs are due to enhanced uptake possibilities. It would be of considerably ecological significance to understand the response of biota to microplastic contamination both physically and physiologically. Here, we report on an area choice experiment (avoidance test) using *Enchytraeus crypticus*, in which we mixed different amounts of high-density polyethylene microplastic particles into the soil. In all experimental scenarios, more Enchytraeids moved to the unspiked sections or chose a lower MP-concentration. Worms in contact with MP exhibited an enhanced oxidative stress status, measured as the induced activity of the antioxidative enzymes catalase and glutathione S-transferase. As plastic polymers per se are nontoxic, the exposure time employed was too short for chemicals to leach from the microplastic, and as the microplastic particles used in these experiments were too large (4 mm) to be consumed by the Enchytraeids, the likely cause for the avoidance and oxidative stress could be linked to altered soil properties.

Keywords: microplastic; *Enchytraeus crypticus*; enchytraeids; avoidance test; toxicity; oxidative stress; catalase; glutathione S-transferase

1. Introduction

The contamination of terrestrial ecosystems and aquatic water bodies with plastics debris has become the so-called chemical footprint of our society. The European MSFD Working Group on Good Environmental Status (WG-GES) classifies plastic pollution as macroplastics (>25 mm), mesoplastics (5 to 25 mm), large microplastics (1 to 5 mm), and small microplastics below the 1 mm size [1]. Nevertheless, the particles do not remain static within these classification brackets, and due to weathering effects and mechanical actions, plastics will continue to degrade into microplastics (MPs) and further, as degradation does not reach a static end-point [2]. The smaller the particles are, the higher the uptake possibilities in organisms, plausibly even allowing the crossing of membranes [3].

Terrestrial ecosystems are on the forefront of MP contamination and are affected conceivably earlier than aquatic ecosystems [4,5]. Our cultivation industry not only uses plastic in the fields, but sludge from wastewater treatment plants that collect MP is used as fertilisers. Changes in soil structure and terrestrial geochemistry (water holding capacity, hydraulic conductivity, soil aggregation, and microbial activity) due to MP pollution have been demonstrated [4,5] which could, in turn, affect species distribution. This creates a toxic environment for the resident biota, which most concerns worms, which are essential in soil turnover and fertilisation. Other reported effects of MP in biota include internal damage due to the consumption or leaching of the additives contained in the plastics [6,7]. Amongst these additives are, e.g., bio-stabilisers, antimicrobials, antioxidants, antistatic agents, blowing agents, fillers/extenders, flame-retardants, fragrances, heat stabilisers, light stabilisers, pigments, and process aids [7]. The leached additives can accumulate in the soil, water, sediment, food, or even body tissues [8]. This could result in an ecosystem that causes severe adverse effects in the native biota. It would be of ecological importance, therefore, to understand if MP pollution could have an influence on the distribution of biota in an ecosystem, as this will contribute to the environmental crisis of decreasing biodiversity. If worm populations would avoid contaminated soils, this, in turn, would affect and alter the soil structure. Therefore, we investigated whether the distribution of biota could be affected by MP pollution in causing mortality or by migration. We selected the oligochaete *Enchytraeus crypticus* as a model organism, due to its abundance in soils globally, and as a likely candidate to be affected by the ubiquitous MP pollution. Enchytraeids are often used as model indicator organisms [9] for various kinds of chemical stressors in terrestrial ecosystems such as lindane, heavy metals, or phenmediphan [10–12]. The oligochaete *E. crypticus* was previously used to estimate the role of pH and PCB No 52 (2,2′,5,5′-Tetrachlorobiphenyl) as well as the effect of soil from former irrigation fields [13,14].

In the present study, mortality tests (using 0%, 2%, 4%, and 8% MP (*w*/*w*)) and avoidance tests (area choice test) were set up. For the avoidance tests, in each case, two choices of either soil void of MP or spiked with 2%, 4%, or 8% of MP (six combinations in total) were presented. The worms could move freely between the soil in two sections of the exposure vessels. To assess the responses of the worms to MP contamination in their environment, we used high-density polyethylene, as it is one of the most widespread plastic materials used today [15]. The MPs (4 mm particles) used were high-density polyethylene (HDPE) (confirmed by Fourier-transform infrared (FTIR) and Raman spectra) produced from threaded bottle caps—common trash seen globally. The MP type, size, and concentration were selected based on monitoring studies which reported on MP pollution in sediments, considering the most commonly. detected MPs and their abundance and size distribution [16,17]. In addition, the size used (4 mm) was explicitly selected by selective sieving, so that the pieces would be too large to be consumed by the Enchytraeids. Thus, we could investigate effects other than consumption. We hypothesised that the worms would avoid MP-contaminated areas or would choose the lower MP concentration of the two options presented. After three days, the number of surviving worms in each section was evaluated, assessing mortality and distribution. Possible adverse effects on *E. crypticus* due to the roaming behaviour at a physiological level was assessed by determining the oxidative stress status measuring catalase and glutathione S-transferase activity as indicators.

2. Materials and Methods

2.1. Enchytraeus Crypticus Culture

Enchytraeus crypticus was continuously cultured at the University of Helsinki under the conditions outlined by Kobeticova et al. [18] and Castro-Ferreira et al. [9]. Briefly, the permanent culture of *E. crypticus* was maintained in the commercially available turf-free soil substrate (MeinWoody, Grub am Forst, Germany), pH 6–7, at a temperature of 18 ± 2 °C. The cultures were fed with oatmeal once a week by mixing the food into the soil substrate. Adults with a well-developed clitellum were used for the tests.

2.2. Microplastic

Plastic from new threaded bottle caps (green colour only) was used for all experiments. However, only caps with the Resin Identification Code (RIC) No. 2 or 02, indicating high-density polyethylene (HDPE)—one of the two most commonly used polymer types [19]—were selected. The caps were washed with tap water to remove possible adherent dirt or dust particles and dried at room temperature before shredding into MP. A desktop plastic recycler (SHR3D IT, 3devo B.V. Utrecht, Netherlands) with a sieve size of 4 mm was used to prepare MP granulate from threated bottle caps. To reach a homogenous granulates size of precisely 4 mm, the material was applied to the shredder five times and then sieved with a series of sieves (Test Sieve ASTM E11 containing steel oven wire) with a different mesh sizes to retain only the 4.00 mm particles (Endecotts Ltd, London, UK) (ISO 3310) and to achieve a homogenous MP sample material. During all stages, caution was taken not to self-contaminate the experimental set-up with other MP particles [20].

Confirmation of the type of the plastic from the bottle caps used was performed using Fourier-transform infrared spectroscopy (FTIR) on a PerkinElmer, Spectrum One (ATR-unit) for IR-spectra, using eight scans with a resolution of 4 cm^{-1} in a range of 4000–650 cm^{-1}. Raman spectroscopy was applied as well using a Renishaw Invia Qontor confocal microscope at 785 nm, grating 1200 l/mm, exposure time 1s, 30 accumulations and 100% laser power, centre 1300 Raman shift/cm^{-1} and a 50-times objective.

2.3. Experimental Set-Up

The same turf-free soil (MeinWoody, Grub am Forst, Germany) used for cultivation was used for experimentation and consisted of 20% lingo fibres, 35% cocopeat washed, 10% spelt fermented, and 35% substrate compost. The soil had a pH ranging between 6 and 7 and was watered to 60% water holding capacity and kept at 18 ± 2 °C for a week.

For the avoidance experiments, a modified protocol based on that described by Amorim et al. [21] was followed. Round paperboard forms with a diameter of 180 mm and a height of 35 mm were used as exposure vessels. To assess acute toxicity, the control vessels were filled with 600 g soil containing no MP (0%) or the respective MP concentration 2%, 4% or 8% mixed as homogeneously as possible (Figure 1). The exposure vessels were divided with a durable paper, adapted to the shape of the vessel, into two parts. For the avoidance tests, one-half was filled with 300 g of MP-free soil (0%) or either 2%, or 4% and the other half was filled with 300 g soil containing the respective MP percentage (2%, 4%, or 8%) (Figure 1).

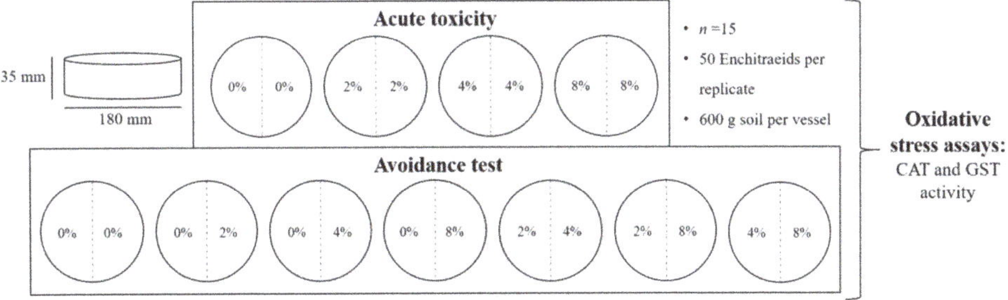

Figure 1. A schematic representation of the experimental setup. CAT: catalase; GST: glutathione S-transferase.

The dividing paper was then removed, and the worms were placed in the middle of the exposure vessel; i.e., the contact line of both soil sides. The avoidance experiments were conducted for 72 h to allow the worms enough time to roam around and make their choice. As the exposure vessels used were larger than those reported by Amorim et al. [21], the longer exposure time was set based on extrapolation and preliminary tests, which assessed the time needed for the worms to travel the longer distances. When terminating the experiment, the soil was separated at the contact using a metal spatula, and the living worms were counted in the separated soil samples. The counted worms were shock-frozen in liquid nitrogen and stored at −80 °C until the extraction of the antioxidative stress enzymes. Mortality was assessed after 72 h of exposure to 0%, 2%, 4%, and 8% MP to assess acute toxicity.

2.4. Oxidative Stress

The worms from the avoidance tests were analysed to assess their oxidative stress status. Enzyme extracts were prepared by homogenising the worms in 0.1 M potassium phosphate buffer pH 6.5 containing 2.17 M glycerol, 1 mM ethylene-diamine-tetra acetic acid (EDTA), and 1.4 mM dithioerythritol (DTE). Cell debris was removed by centrifugation (10 min at 13,000 × g), and the supernatant was used for enzyme measurements [22]. Catalase activity (CAT, E.C. 1.11.1.6) was measured on a Tecan Infinite F Nano+ plate reader at 240 nm with the decrease of absorbance correlating to the disappearance of H_2O_2 [23]. The reaction mixture contained 50 mmol/L sodium phosphate buffer, 10 mmol/L H_2O_2 and 10 μL enzyme extract. The enzyme activity of CAT was defined as 1 mmol of H_2O_2 oxidised over 1 min at 25 °C and expressed as μkat/mg protein. Soluble (cytosolic) glutathione S-transferases (E.C. 2.5.1.18) were determined using the standard model substrate 1-chloro-2,4-dinitrobenzene (CDNB), according to Habig et al. [24]. Enzyme activities are given in katal per milligram protein (kat/mg protein); where katal (kat) is the conversion rate of one mol substrate per second. The protein content of each sample was determined according to the method of Bradford [25] using the Bradford protein-dye reagent (Sigma). Bovine serum albumin (98%, Sigma.Aldrich, St. Louis, Missouri, United States) was used as a standard protein for calibration of the assay method.

2.5. Statistical Analysis

SPSS software (IBM SPSS Statistics, Version 20) was used to perform a descriptive analysis based on the mean of the different endpoints chosen. Normality and homogeneity of variance were tested using Shapiro–Wilk and Levene's test, respectively. When data proved to be normally and homogenously distributed, the data were submitted to the one-way analysis of variance (ANOVA) followed by Tukey post hoc in SPSS. When tests for normality and homogeneity were not satisfied, the non-parametric Mann–Whitney-U Test together with pairwise comparisons was employed. We set the alpha value as 0.05 level for significance [26]. All data are graphically expressed as mean ± standard deviation (SD).

3. Results and Discussion

3.1. Toxicity and Avoidance Tests

An avoidance test is defined as an organism's active selection between two samples exhibiting different properties [27]. Therefore, in the present study, *Enchytraeus* sp. worms were placed in exposure vessels containing two non-obstructed halves, which consisted of soil mixed with different percentages of MP ranging from 0% to 8% (*w*/*w*) to determine their preference, if any. FTIR and Raman spectroscopic analyses of the bottle cap plastics confirmed that they were indeed HDPE (Figure A1). To assess the acute toxicity (mortality), the two halves consisted of the same MP percentages; i.e., with no MP on both sides, or 2%, 4%, or 8% on both sides, respectively. In all these mortality tests, where the exposure percentages were equal throughout the exposure vessel, *E. crypticus* distributed equally between the two halves (Table 1).

Table 1. Acute toxicity and area preference in pairings with the same percentage of microplastic (MP) in both halves (n = 15, at 50 worms per independent replicate). Significance was tested by pairwise t-tests after normality and homogeneity tests were satisfied.

%MP in Soil Halves (*w*/*w*)	Worms in	Worms out	Mortality	Distribution in the Two Exposure Vessel Halves		Distribution Comparison (*P*-Value)
0%	50 ± 0	49 ± 1	2%	25 ± 3	24 ± 3	0.328
2%	50 ± 0	46 ± 2	8%	23 ± 2	23 ± 3	0.618
4%	50 ± 0	45 ± 1	10%	23 ± 3	22 ± 2	0.186
8%	50 ± 0	43 ± 1	14%	22 ± 2	21 ± 2	0.340

An increased percentage of HDPE MP in the soil (0% to 8%) resulted in the *E. crypticus* mortality increasing from 2% to 14% (Table 1). However, the properties of the monomer ethylene used to produce HDPE were previously reported not to cause toxicity nor to exhibit estrogenic activity [28]. Consumption leading to internal obstruction and damage is unlikely due to the size of the particles used. However, due to the manufacturing process, all plastics can contain residual chemicals, including catalysts necessary for the polymerisation reactions, which could quickly leach from new plastics. Additives such as stabilisers, UV-blockers, plasticisers, antioxidants, and colourants are added to the plastic formulation to provide the final product with the necessary properties. These additives are retained in the plastic bound to the polymer matrix through van der Waals forces [29]. The leaching of those chemicals due to the breakage of these weak bonds during the degradation of plastics might therefore occur [7,30] and affect our environment [28,31]. The toxicity observed here could be due to leaching additives from the shredded bottle caps [4,5]. However, it is more likely that the MP particles could have caused changes in the soil structure [4,5], which resulted in undesirable conditions for the Enchytraeids [21].

In all avoidance test pairings, where non-spiked soil was presented against MP spiked soil (0% to 2%, 0% to 4%, and 0% to 8%), more *E. crypticus* (avg. 60% ± 4%) moved to the non-spiked half (Figure 2). The Enchitraeids' preference was higher by factors of 1.6 to 1.8 in favour of the non-spiked side. The average survival rates in the pairings with an unspiked side ranged from 80% to 96%. Following the experimental set-up from Kerekes and Feigl [32], all MP concentrations were paired against each other for the avoidance tests; i.e., 2% to 4%, 2% to 8%, and 4% to 8%. In these pairings offering a lower and a higher MP concentration as a choice, *E. crypticus* also preferred the lower MP concentration in all pairings (Figure 2). The worms favoured the lower MP concentration side by factors of 1.7 to 2.7 with increasing MP percentages. Thus, in the avoidance tests, Enchytraeids showed a clear preference for the MP-free sides or less polluted sides, most likely due to altered soil properties, such as decreased bulk soil density and decreased microbial activity [5] as the MP particles were too large to consume. The possibility of leaching cannot be completely excluded as a potential reason for the avoidance; however, this is unlikely as the exposure time was too short for leaching and exposure was carried out at room temperature (18 ± 2 °C) and not under solar irradiation [33–35].

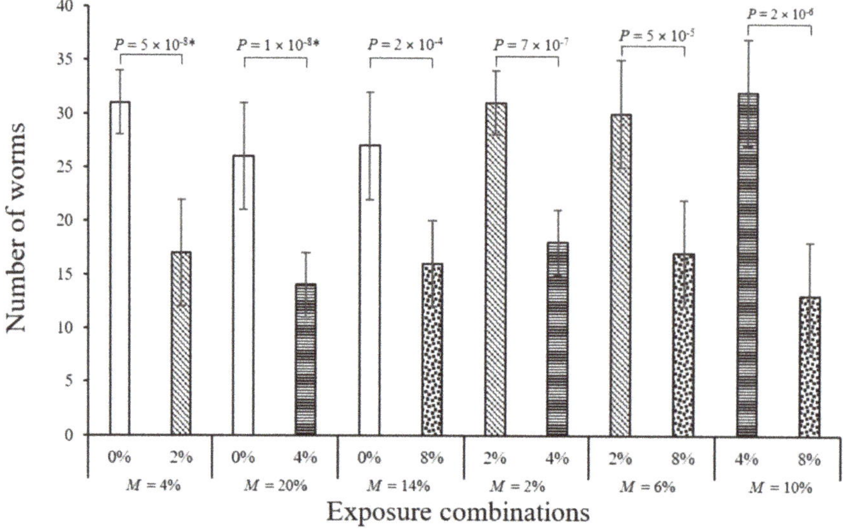

Figure 2. Worms counted in non-spiked and MP-spiked areas of the avoidance test vessel. Data represent mean worm count per area ± standard deviation (n = 15, at 50 worms per independent replicate). The average percentage mortality per pairing is given under each section as *M*. Differences between the treatments were tested by one-way ANOVA and Tukey post hoc when the data were normally and homogeneously distributed. When data were not homogenously distributed, even after transformation, the non-parametric Kruskal–Wallis test with pairwise comparisons was used.

3.2. Oxidative Stress

As exposure to MP is correlated to oxidative stress [36], we measured the activity of catalase (CAT) and glutathione S-transferase (GST) in the worms after exposure to different HDPE MP percentages in soil. Exposure to the HDPE MP caused the CAT activity of the Enchytraeids to increase dose-dependently (Figure 3A). In the pairings consisting of 0% to 8% as well as 2% to 8% MP, the worms exhibited higher CAT activity (Figure 3C), suggesting that 8% (*w*/*w*) MP in soil induced oxidative stress in these worms.

GST is known as a biotransformation enzyme; however, it is also involved in the antioxidative stress defence as it metabolises end-products such as malondialdehyde and 4-hydroxynonenal derived from lipid-peroxidation [22]. As with CAT, the increasing HDPE MP percentages in the soil resulted in a dose-dependent increase of the GST enzyme activity in the Enchytraeids (Figure 3B). All pairings, except 0% to 2% HDPE MP, resulted in elevated GST activity (Figure 3D).

In most of the cases presented here, exposure to MP in the soil led to an increase in enzyme activity, indicating the elicitation of an antioxidative stress response. For nanoparticles and microbeads with a size range between 0.05 and 6 μm, it is known that the toxicity is closely correlated to the uptake into organisms and the generation of reactive oxygen species (ROS) [36–38]. An increase of ROS will lead to oxidative-stress-induced signalling pathways. However, the MP particles used here were specifically selected to have a size of 4 mm; therefore, they are too large to be taken up by the oligochaetes used. As leaching is implausible, altered soil properties may have induced oxidative stress in agreement with the findings reported by Howcroft et al. [39].

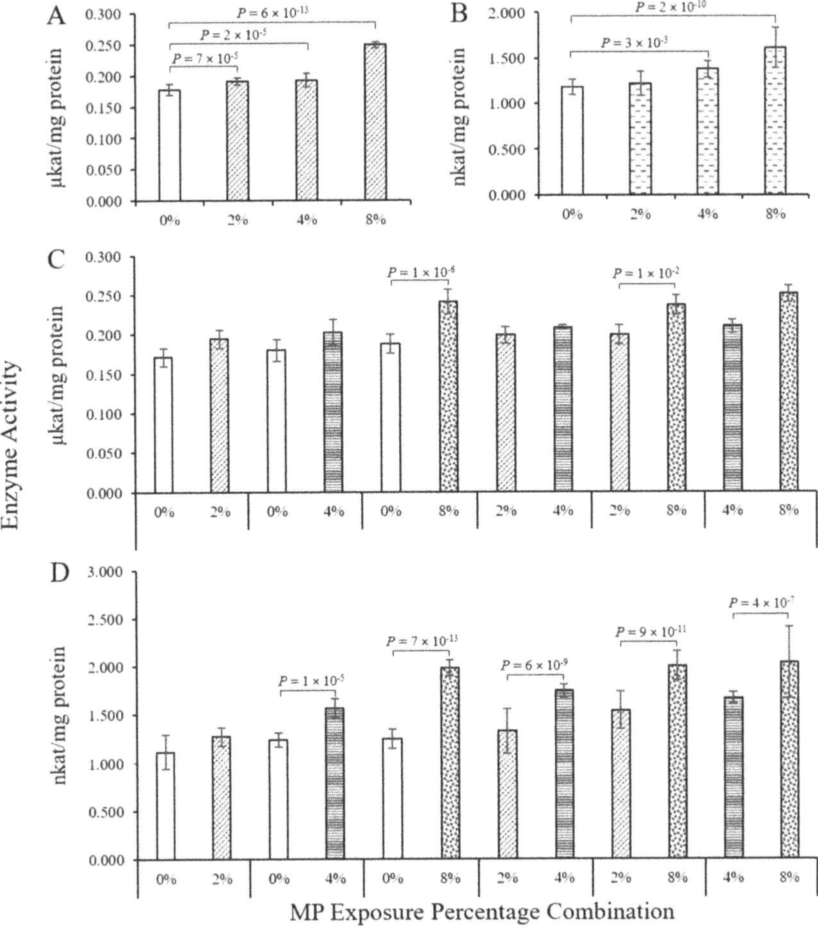

Figure 3. (**A**) Catalase activity in the 0% MP on both sides pairing, as well as in pairings with the same MP concentration on both sides. (**B**) GST activity in a 0% to 0% pairing, as well as in pairings with the same MP percentage on both sides. Data represent mean enzyme activity ± standard deviation ($n = 9$, at 50 worms per independent replicate). (**C**) Catalase activity in worms from soil containing 0% to 2%; 4%, and 8% MP, as well as worms from the avoidance experiment from clean and MP-spiked sides of different MP concentrations. (**D**) Glutathione S-transferase activity in control worms from soil containing 2%, 4% and 8% MP, as well as worms from the avoidance experiment from clean and MP-spiked sides of different MP concentrations. Data represent mean enzyme activity ± standard deviation ($n = 3$, at 50 worms per independent replicate). When data were not homogenously distributed, even after transformation, the non-parametric Kruskal–Wallis test with pairwise comparisons were used.

In conclusion, the results show that the oligochaetes preferred an MP-free environment and that, in the presence of MP, their antioxidative stress response was elevated. As uptake and leaching under the experimental conditions used here are unlikely, altered soil properties due to the presence of MP may be the cause for the results observed. More research is needed to investigate long-time exposure and the toxicity of the compounds leaching from MP in our environment to better understand the adverse effects of MP in our ecosystems. This is the first study to show an area choice test for Enchytraeids avoiding MP spiked sites.

Author Contributions: S.P., M.E.L., S.M.M., S.K., Y.J.K., O.-P.P. conceived the study, S.P. and J.H.H. collected the data, S.P. and M.E.-L. analysed the data and prepared the figures, S.P. and M.E.-L. wrote the first draft, A.-M.v.W., O.-P.P., Y.J.K., S.K., and S.M.M. reviewed and amended the paper. All authors have read and agreed to the published version of the manuscript.

Funding: Open access funding provided by University of Helsinki, including Helsinki University Central Hospital.

Acknowledgments: Open access funding provided by University of Helsinki. We thank Saija M. Saami for her critical reading of the manuscript and valuable comments.

Conflicts of Interest: The authors declare no conflict of interest.

Appendix A

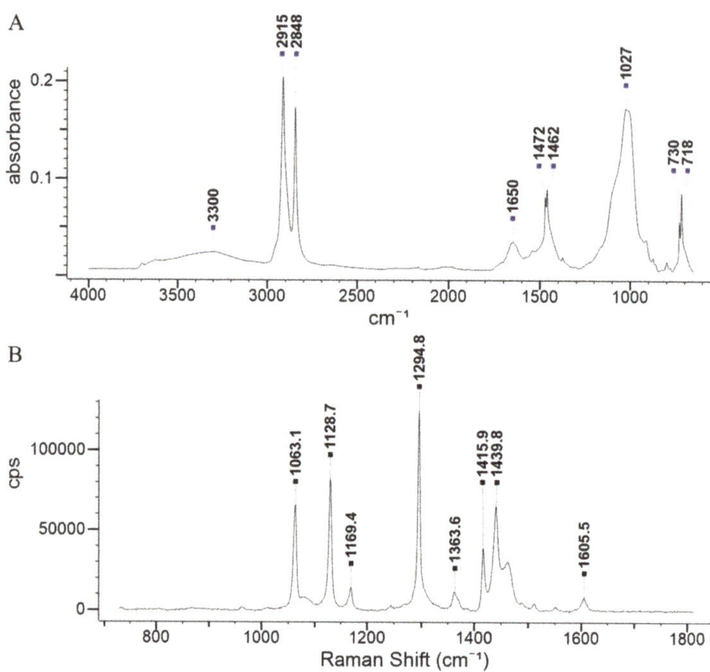

Figure A1. Confirmation of the MP type as high-density polyethylene (HDPE) with (**A**) Fourier transform infrared (FTIR) spectra and (**B**) Raman spectra.

References

1. Hanke, G. *Guidance on Monitoring of Marine Litter in European Seas;* A Guidance Document within the Common Implementation Strategy for the Marine Strategy Framework Directive; J.R.C. MSDF Technical Subgroup on Marine Litter: Luxembourg, 2013; p. 128.
2. Koelmans, A.A. Modelling the role of microplastics in bioaccumulation of organic chemicals to marine aquatic organisms. A critical review. In *Marine Anthropogenic Litter*, 2nd ed.; Bergmann, M., Gutow, L., Klages, M., Eds.; Springer: Berlin, Germany, 2015; pp. 313–328.
3. Ng, E.L.; Lwanga, E.H.; Eldridge, S.M.; Johnston, P.; Hu, H.W.; Geissen, V.; Chen, D. An overview of microplastic and nanoplastic pollution in agroecosystems. *Sci. Total Environ.* **2018**, *627*, 1377–1388. [CrossRef]
4. De Souza Machado, A.A.; Kloas, W.; Zarfl, C.; Hempel, S.; Rillig, M.C. Microplastics as an emerging threat to terrestrial ecosystems. *Glob. Chang. Biol.* **2018**, *24*, 1405–1416. [CrossRef] [PubMed]
5. De Souza Machado, A.A.; Lau, C.W.; Till, J.; Kloas, W.; Lehmann, A.; Becker, R.; Rillig, M.C. Impacts of microplastics on the soil biophysical environment. *Environ. Sci. Technol.* **2018**, *52*, 9656–9665. [CrossRef]

6. Lithner, D.; Larsson, A.; Dave, G. Environmental and health ranking and assessment of plastic polymers based on chemical composition. *Sci. Total Environ.* **2011**, *409*, 3309–3324. [CrossRef] [PubMed]
7. Hermabessiere, L.; Dehaut, A.; Paul-Pont, I.; Lacroix, C.; Jezequel, R.; Soudant, P.; Duflos, G. Occurrence and effects of plastic additives on marine environments and organisms: A review. *Chemosphere* **2017**, *182*, 781–793. [CrossRef] [PubMed]
8. Crompton, T. Additive Migration from Plastics into Foods. In *A Guide for Analytical Chemistry*, 1st ed.; Smithers Rapra Technology Publishing: Shrewsbury, UK, 2007.
9. Castro-Ferreira, M.P.; Roelofs, D.; van Gestel, C.A.M.; Verweij, R.A.; Soares, A.M.V.M.; Amorim, M.J.B. *Enchytraeus crypticus* as model species in soil ecotoxicology. *Chemosphere* **2012**, *87*, 1222–1227. [CrossRef]
10. Didden, W.A.M.; Römbke, J. Enchytraeids as indicator organisms for chemical stress in terrestrial ecosystems. *Ecotoxicol. Environ. Saf.* **2001**, *50*, 25–43. [CrossRef]
11. Amorim, M.J.; Sousa, J.P.; Nogueira, A.J.A.; Soares, A.M.V.M. Comparison of chronic toxicity of Lindane (g-HCH) to *Enchytraeus albidus* in two soil types: The influence of soil pH. *Pedobiologia* **1999**, *43*, 635–640.
12. Amorim, M.J.; Sousa, J.P.; Nogueira, A.J.A.; Soares, A.M.V.M. Bioavailability and toxicokinetics of C-14-lindane (gamma-HCH) in the enchytraeid *Enchytraeus albidus* in two soil types: The aging effect. *Arch. Environ. Contam. Toxicol.* **2002**, *43*, 221–228. [CrossRef]
13. Achazi, R.K.; Chroszcz, G.; Pilz, B.; Rothe, B.; Steudel, I.; Throl, C. Der Einfluss des pH-Werts und von PCB52 auf Reproduktion und Besiedlungsaktivität von terrestrischen Enchytraeen in PAK-, PCB- und schwermetallbelasteten Rieselfeldböden. *Verh. Ges. Ökol.* **1996**, *26*, 37–42.
14. Achazi, R.K.; Fröhlich, E.; Henneken, M.; Pilz, C. The effect of soil from former irrigation fields and of sewage sludge on dispersal activity and colonizing success of the annelid *Enchytraeus crypticus* (Enchytraeidae, Oligochaeta). *Newslett. Enchytraeidae* **1999**, *6*, 117–126.
15. Shah, A.A.; Hasan, F.; Hameed, A.; Ahmed, S. Biological degradation of plastics: A comprehensive review. *Biotechnol. Adv.* **2008**, *26*, 246–265. [CrossRef] [PubMed]
16. Haave, M.; Lorenz, C.; Primpke, S.; Gerdts, G. Different stories told by small and large microplastics in sediment - first report of microplastic concentrations in an urban recipient in Norway. *Mar. Pollut. Bull.* **2019**, *141*, 501–513. [CrossRef] [PubMed]
17. Scopetani, C.; Chelazzi, D.; Cincinelli, A.; Esterhuizen-Londt, M. Assessment of microplastic pollution: Occurrence and characterisation in Vesijarvi lake and Pikku Vesijarvi pond, Finland. *Environ. Monit. Assess.* **2019**, *191*, 652. [CrossRef]
18. Kobeticova, K.; Hofman, J.; Holoubek, I. Ecotoxicity of wastes in avoidance tests with *Enchytraeus albidus*, *Enchytraeus crypticus* and *Eisenia fetida* (Oligochaeta). *Waste Manag.* **2010**, *30*, 558–564. [CrossRef]
19. Arutchelvi, J.; Sudhakar, M.; Arkatkar, A.; Doble, M.; Bhaduri, S.; Uppara, P.V. Biodegradation of polyethylene and polypropylene. *Indian J. Biotechnol.* **2008**, *7*, 9–22.
20. Scopetani, C.; Esterhuizen-Londt, M.; Chelazzi, D.; Cincinelli, A.; Setälä, H.; Pflugmacher, S. Self-contamination from clothing in microplastics research. *Ecotoxicol. Environ. Saf.* **2020**, *189*, 110036. [CrossRef]
21. Amorim, M.J.B.; Novais, S.; Römbke, J.; Soares, A.M.V.M. Avoidance test with *Enchytraeus albidus* (Enchytraeidae): Effects of different exposure time and soil properties. *Environ. Pollut.* **2008**, *155*, 112–116. [CrossRef]
22. Pflugmacher, S. Promotion of oxidative stress in *C. demersum* due to exposure to cyanobacterial toxin. *Aquat. Toxicol.* **2004**, *3*, 169–178. [CrossRef]
23. Aebi, H. Catalase in vitro. *Methods Enzymol.* **1984**, *105*, 121–126.
24. Habig, W.; Pabst, M.J.; Jacoby, W.B. Glutathione-S-transferases. *J. Biol. Chem.* **1974**, *249*, 1730–1739.
25. Bradford, M.M. A rapid and sensitive method for the quantitation of microgram quantities of protein utilizing the principle of protein-dye binding. *Anal. Biochem.* **1976**, *72*, 248–254. [CrossRef]
26. Sokal, R.R.; Rohlf, F.J. *Biometry: The Principles and Practice of Statistics in Biological Research*; WH Freeman and Company: New York, NY, USA, 1997.
27. Coleman, D.C.; Wall, D.H. Chapter 5—Soil Fauna: Occurrence, biodiversity and roles in ecosystem function. In *Soil Microbiology, Ecology and Biochemistry*, 4th ed.; Paul, E.A., Ed.; Academic Press: Boston, MA, USA, 2015; pp. 111–149.
28. Yang, C.Z.; Yaniger, S.I.; Jordan, V.C.; Klein, D.J.; Bittner, G.D. Most plastic products release estrogenic chemicals: A potential health problem that can be solved. *Environ. Health Perspect.* **2011**, *119*, 989–996. [CrossRef] [PubMed]
29. Zhang, X.; Chen, Z. Observing phthalate leaching from plasticized polymer films at the molecular level. *Langmuir* **2014**, *30*, 4933e4944. [CrossRef] [PubMed]

30. Suhrhoff, T.J.; Scholz-Böttcher, B.M. Qualitative impact of salinity, UV radiation and turbulence on leaching of organic plastic additives from four common plastics—A lab experiment. *Mar. Pollut. Bull.* **2016**, *102*, 84–94. [CrossRef]
31. Stern, B.R.; Lagos, G. Are there health risks from the migration of chemical substances from plastic pipes into drinking water? A Review. *Hum. Ecol. Risk Assess.* **2008**, *14*, 753–779. [CrossRef]
32. Kerekes, I.K.; Feigl, V. The effect of bauxite residues on the avoidance behaviour of *Enchytraeus albidus* (Enchytraeidae). *Period. Polytech. Chem. Eng.* **2018**, *62*, 415–425. [CrossRef]
33. Loyo-Rosales, J.E.; Rosales-Rivera, G.C.; Lynch, A.M.; Rice, C.P.; Torrents, A. Migration of nonylphenol from plastic containers to water and a milk surrogate. *J. Agric. Food Chem.* **2004**, *52*, 2016–2020. [CrossRef]
34. Romera-Castillo, C.; Pinto, M.; Langer, T.M.; Álvarez-Salgado, X.A.; Herndl, G.J. Dissolved organic carbon leaching from plastics stimulates microbial activity in the ocean. *Nat. Commun.* **2018**, *9*, 1430. [CrossRef]
35. Chen, Q.; Allgeier, A.; Yin, D.; Hollert, H. Leaching of endocrine disrupting chemicals from marine microplastics and mesoplastics under common life stress conditions. *Environ. Int.* **2019**, *130*, 104938. [CrossRef]
36. Jeong, C.B.; Kang, H.M.; Lee, M.C.; Kim, D.H.; Han, J.; Hwang, D.S.; Souissi, S.; Lee, S.J.; Shin, K.H.; Park, H.G.; et al. Adverse effects of microplastics and oxidative stress-induced MAPK/Nrf2 pathway-mediated defense mechanisms in the marine copepod *Paracyclopina nana*. *Sci. Rep.* **2017**, *7*, 41323. [CrossRef] [PubMed]
37. Okupnik, A.; Pflugmacher, S. Oxidative stress response of the aquatic macrophyte *Hydrilla verticillata* exposed to TiO_2 nanoparticles. *Environ. Toxicol. Chem.* **2016**, *35*, 2859–2866. [CrossRef]
38. Spengler, A.; Wanninger, L.; Pflugmacher, S. Oxidative stress mediated toxicity of TiO_2 nanoparticles after a concentration and time dependent exposure of the aquatic macrophyte *Hydrilla verticillata*. *Aquat. Toxicol.* **2017**, *190*, 32–39. [CrossRef] [PubMed]
39. Howcroft, H.; Amorim, M.J.B.; Gravato, C.; Guilhermino, L.; Soares, A.M.V.M. Effects of natural and chemical stressors on Enchytraeus albidus: Can oxidative stress parameters be used as fast screening tools for the assessment of different stress impacts in soils? *Environ. Int.* **2009**, *35*, 318–324. [CrossRef] [PubMed]

MDPI
St. Alban-Anlage 66
4052 Basel
Switzerland
Tel. +41 61 683 77 34
Fax +41 61 302 89 18
www.mdpi.com

Toxics Editorial Office
E-mail: toxics@mdpi.com
www.mdpi.com/journal/toxics

www.ingramcontent.com/pod-product-compliance
Lightning Source LLC
LaVergne TN
LVHW070610170726
843515LV00004B/36

* 9 7 8 3 0 3 6 5 1 2 9 7 6 *